Lecture Notes
in Economics and
Mathematical Systems

Managing Editors: M. Beckmann and H. P. Künzi

183

Klaus Schittkowski

Nonlinear Programming Codes

Information, Tests, Performance

Springer-Verlag
Berlin Heidelberg New York 1980

Author

Klaus Schittkowski
Institut für Angewandte
Mathematik und Statistik
Universität Würzburg
Am Hubland
8700 Würzburg
Federal Republic of Germany

AMS Subject Classifications (1980): 90-04, 90 C 30

ISBN-13: 978-3-540-10247-2   e-ISBN-13: 978-3-642-46424-9

DOI: 10.1007/978-3-642-46424-9

2141/3140-543210

Preface ....

The increasing importance of mathematical programming for the
solution of complex nonlinear systems arising in practical situations
requires the development of qualified optimization software. In
recent years, a lot of effort has been made to implement efficient and
reliable optimization programs and we can observe a wide distribution
of these programs both for research and industrial applications. In
spite of their practical importance only a few attempts have been made
in the past to come to comparative conclusions and to give a designer
the possibility to decide which optimization program could solve his
individual problems in the most desirable way.

Box [BO 1966], Huang, Levy [HL 1970], Himmelblau [HI 1971], Dumitru [DU 1974], and Moré, Garbow, Hillstrom [MG 1978] for example
compared algorithms for unrestricted optimization problems. Bard
[BD 1970], McKeown [MK 1975], and Ramsin, Wedin [RW 1977] studied
codes for nonlinear least squares problems. Codes for the linear case
are compared by Bartels [BA 1975] and Schittkowski, Stoer [SS 1979].
Extensive tests for geometric programming algorithms are found in
Dembo [DE 1976b], Rijckaert [RI 1977], and Rijckaert, Martens [RM 1978].
For the general case, i.e. minimization of an arbitrary objective
function under constraints, we have to mention first the Colville
[CO 1968] report. Other relevant papers are Stocker [ST 1969],
Holzman [HO 1969], Tabak [TA 1969], Beltrami [BE 1969], Miele, Tietze,
Levy [MT 1972], Eason, Fenton [EF 1972], Asaadi [AS 1973], Staha
[SH 1973], and Sandgren [SA 1977].

.Comparative studies of the latter kind are either designed to test
special modifications of a mathematical algorithm or to perform
general purpose software tests. The number of optimization programs
varies from 2 to 30 executed in up to 35 different versions. Most
programs have been realizations of penalty methods and are therefore
antiquated from the current point of view. In particular, none of
the comparative studies tested so called quadratic approximation
or recursive quadratic programming methods which will find

growing interest both for future research efforts and applications
because of their outstanding efficiency. In this study, we will intro-
duce 20 different optimization codes in 26 versions. Additional versions
are used to test the effect of numerical differentiation. Among these
programs, we find realizations of quadratic approximation, generalized
reduced gradient, multiplier, and penalty methods. Most of them are
currently used to solve practical nonlinear programming problems in
various kinds of applications. It is one of the intensions of this
report to give technical information about the programs such as source,
language, length, provision of problem functions, etc.. The programs
will be tested extensively from different points of view to give a
user the possibility to choose the most appropriate one for solving
his individual optimization problems.

The exploitation of any experimental tests of nonlinear programming
software depends on the quality of the used test problems since all
conclusions can be confirmed only by numerical experiments. The test
examples of earlier comparative studies are composed of small, arti-
ficial or, more often, of so called 'real life' problems which are
believed to reflect typical structures of practical nonlinear programming problems, see Himmelblau [HI 1972], Cornwell, Hutchison,
Minkoff, Schultz [CH 1978], or Hock, Schittkowski [HS 1980], but
these test examples have some severe disadvantages especially since
the precise solution is not known a priori. Therefore it is not possible
to evaluate the achieved accuracy of an optimization program and to
relate the efficiency, i.e. calculation time, number of function and
gradient evaluations, to the accuracy. Furthermore, one has in general
too little information on the mathematical structure of the test
problems so that in the past, the efficiency of an optimization
program was determined mainly in terms of calculation time or of the
number of function and gradient evaluations. These drawbacks gave the
impulse to develop a test problem generator which is capable to compute
test problems randomly with predetermined (at least) local solutions.
We are able now to determine not only the efficiency of a program and
to relate it to the achieved accuracy, but also to develop quantitative
measures for reliability and global convergence. Furthermore it is
possible to generate convex, ill-conditioned, degenerate, and indefinite problems which are used for special purpose evaluations and to
check the performance of optimization programs in extreme situations.
To satisfy all these conditions, we generated 185 test problems randomly

with predetermined solutions. Since most of them are executed with different starting points, each optimization code under consideration has to pass 370 test runs in contrast to at most 30 test runs performed in earlier studies.

A reader who is interested in selecting a program for the numerical solution of his optimization problems, could use the following guiding rules: He should start with Chapter I where the problem is formulated and then proceed to Chapter III. The first section will give him a survey of technical details such as length, provision of problem functions, embedded numerical differentiation, etc., leading to a special subset of programs which satisfy certain technical assumptions and which could be implemented to solve his problems. More detailed information and a rough sketch of their performance are contained in the second section where all programs are described individually. Subsequently, the decision maker should read Chapter V where the following nine performance criteria are evaluated:

> Efficiency.
> Reliability.
> Global convergence.
> Performance for solving degenerate problems.
> Performance for solving ill-conditioned problems.
> Performance for solving indefinite problems.
> Sensitivity to slight variations of the problem.
> Sensitivity to the position of the starting point.
> Ease of use.

All these criteria are evaluated in a quantitative manner and, in addition, it is outlined how a final score may be obtained for optimization programs according to the individual significance of the criteria for the decision maker. More detailed numerical results and information about the performance evaluation are contained in Appendices C and D allowing a thorough investigation for special problem types.

In addition, the mathematical background of the algorithms is described in Chapter II and Chapter IV shows how test problems with predetermined solutions are generated randomly. Numerical data for their construction in the scope of this comparative study and a

sensitivity analysis are given in Appendices A and B. Final conclusions, recommendations for the design of optimization programs,
and some technical remarks are gathered in Chapter VI.

I would like to express my sincere gratitude to all authors for
submitting their optimization programs and especially to the Rechenzentrum of the University of Würzburg for the support making it
possible to perform the extensive numerical tests. The performance
evaluation was influenced by many fruitful discussions with other
COAL members (Committee on Algorithms of the Mathematical Programming
Society). Furthermore I would like to thank J. Stoer and J. Abadie
for helpful comments and suggestions.

Contents ....

INTRODUCTION

The general model of nonlinear programming is to minimize an arbitrary objective function subject to nonlinear equality and inequality constraints. The mathematical formulation is the following one:

$$\min \quad f(x)$$

$$g_j(x) = 0 \ , \quad j=1,\ldots,m_e$$

$$x \in \mathbb{R}^n: \quad g_j(x) \geq 0 \ , \quad j=m_e+1,\ldots,m \qquad (1)$$

$$x_1 \leq x \leq x_u$$

The functions $f, g_j: \mathbb{R}^n \to \mathbb{R}$, $j=1,\ldots,m$, are assumed to be continuously differentiable. Without loss of generality, we include lower and upper bounds x_1 and x_u, respectively.

In the following, the symbols $\nabla_x h$ and $\nabla_x^2 h$ represent the first and second derivatives of a function h with respect to x, i.e. if $h: \mathbb{R}^n \to \mathbb{R}$ is twice differentiable and $x \in \mathbb{R}^n$, then

$$\nabla_x h(x) := (\frac{\delta}{\delta x_1} h(x), \ \ldots \ , \frac{\delta}{\delta x_n} h(x))^T$$

denotes the transpose of the gradient of h in x and

$$\nabla_x^2 h(x) := \left(\frac{\delta^2}{\delta x_i \delta x_j} h(x) \right)_{i,j=1,n}$$

the Hessian matrix of h in x. In addition, we denote by $D_x H(x)$ the Jacobian matrix of a function $H: \mathbb{R}^n \to \mathbb{R}^m$, i.e. if $x \in \mathbb{R}^n$ and $H(x) := (H_1(x),\ldots,H_m(x))^T$, then

$$D_x H(x) := \left(\frac{\delta}{\delta x_j} H_i(x) \right)_{i=1,m; j=1,n} \quad .$$

Of fundamental importance both for nonlinear programming theory and for developing optimization algorithms is the Lagrange function

$$L(x,u) := f(x) - \sum_{j=1}^{m} u_j \, g_j(x) \qquad (2)$$

defined for all $x \in \mathbb{R}^n$ and $u = (u_1,\ldots,u_m)^T \in \mathbb{R}^m$. Since in the following we are mainly interested in sufficient optimality conditions, we summarize the main properties of the Lagrangian (2) in the following theorems:

Theorem 1: Let $f,g_1,\ldots,g_m$ be twice differentiable functions. A point $x^* \in \mathbb{R}^n$ with $x_1 < x^* < x_u$ is an isolated local minimizer of (1), if there exists a vector $u^* = (u_1^*,\ldots,u_m^*)^T$, such that the following conditions are valid:

a) Kuhn-Tucker condition:

$$g_j(x^*) = 0 , \quad j=1,\ldots,m_e.$$
$$g_j(x^*) \geq 0 , \quad j=m_e+1,\ldots,m.$$
$$u_j^* \geq 0 , \quad j=m_e+1,\ldots,m. \qquad (3)$$
$$u_j^* \, g_j(x^*) = 0 , \quad j=m_e+1,\ldots,m.$$
$$\nabla_x L(x^*,u^*) = 0$$

b) Second order condition: For every nonzero vector y where $y^T \nabla_x g_j(x^*) = 0$, $j=1,\ldots,m_e$, and $y^T \nabla_x g_j(x^*) = 0$ for all j with $u_j^* > 0$, $j=m_e+1,\ldots,m$, it follows that

$$y^T \nabla_x^2 L(x^*,u^*) y > 0 . \qquad (4)$$

The proof is standard and can be found, e.g., in McCormick [MC 1967]. In addition, it is easy to show the following saddle-point condition:

Theorem 2: If there exists a saddle-point (x^*,u^*) of L, i.e. if $x_1 < x^* < x_u$, $u^* = (u_1^*,\ldots,u_m^*)^T$ with $u_j^* \geq 0$ for $j > m_e$, and

$$L(x^*,u) \leq L(x^*,u^*) \leq L(x,u^*) \qquad (5)$$

for all $x \in \mathbb{R}^n$ and for all $u = (u_1,\ldots,u_m)^T \in \mathbb{R}^m$ with $u_j \geq 0$ for $j > m_e$, then x^* defines a global minimizer of (1).

Both theorems require the existence of a u* ∈ $\mathbb{R}^m$ satisfying special properties. We denote this vector as the vector of the Lagrange multipliers. The subsequent chapter shows that modern optimization algorithms do not only approximate the solution x* but also the Lagrange multipliers u* simultaneously.

OPTIMIZATION METHODS

The optimization programs tested in the scope of this report
realize various mathematical algorithms and the intention of this
chapter is to give a rough sketch of their concepts. In particular,
we describe the fundamental mathematical methods without any attempt
to be complete and give references where the interested reader can
find further details, convergence results, etc.. Since all programs
require auxiliary routines for special types of subproblems, we
begin the chapter with reviews of line-search, quadratic programming,
and unconstrained optimization algorithms. Since these methods are
often ignored in the corresponding user instructions of the constrained
nonlinear programming codes, we explain them more comprehensively and
present implementation details together with some sources of computer
programs. Subsequently, the methods for solving the general constrained
problem (1) which are realized numerically by the submitted optimization programs are described. They consist of penalty, multiplier,
quadratic approximation, generalized reduced gradient algorithms, and
a special method developed by Robinson. More details and some information about their individual implementation are given in Chapter III.

1. Line-search algorithms

Nearly all optimization codes require the computation of a steplength $\bar{\alpha}$ for determining a new iterate $\bar{x}$ of the form

$$\bar{x} := x + \bar{\alpha}d , \qquad (6)$$

where x denotes the current iterate and d the so called search
direction obtained by one of the strategies described in the subsequent sections. To explain the methods for determining $\bar{\alpha}$, let us
proceed from the idea of minimizing an unconstrained real-valued

function F on $\mathbb{R}^n$. The best choice of $\bar{\alpha}$ would minimize F(x) along the search direction. Since we assume that d is a descent direction, the one-dimensional search is restricted to the interval $(0,\infty)$, i.e. we have to solve the subproblem

$$\min \{F(x + \alpha d): 0 < \alpha < \infty\}. \tag{7}$$

The special structure of the function F depends on the underlying mathematical technique for solving the constrained problem (1) and could be given by one of the following definitions:

F(x):= f(x)
- f objective function.

F(x):= P(x,r)
- P penalty function, r penalty parameter.

F(x):= Ψ(x,u,s)
- Ψ augmented Lagrange function, s penalty parameter, u approximation of the Lagrange multipliers.

F(x):= $f(x^b(x^{nb}),x^{nb})$
- f objective function with variables partitioned in so called basis variables $x^b(.)$ and non-basis variables x^{nb}.

Although F could be non-differentiable under certain circumstances, we proceed now from the assumption that F possesses continuous gradients $\nabla F(x)$ for all $x \in \mathbb{R}^n$.

Since there is no chance of obtaining the minimizer $\bar{\alpha}$ of (7) precisely and since any iterative process to compute $\bar{\alpha}$ within machine accuracy would require too many function or derivative evaluations of F(x) in general, we have to look for an adequate approximation, i.e. for a compromise between minimizing the overall costs on the one side and the accuracy on the other side which is required to guarantee convergence. Although most steplength procedures are based on similar concepts, we can observe a wide variety of different implementations in existing optimization programs. Line-search algorithms influence the efficiency of a program significantly, see for example Dixon [DI 1972], and an appropriate implementation does not only depend on the underlying method for solving problem (1), i.e. the function F, but also on the individual problem which has to be solved, especially on the question if the usage of derivatives is permitted.

1.1 Line-search algorithms with guaranteed convergence

The main feature of the considered line-search algorithms is that the steplength does not converge to the minimizer of (7), but under mild conditions on the search direction d, it is possible to prove the convergence of the underlying descent method for minimizing F(x). The following two conditions for determining $\bar{\alpha}$ are well-known because of their simple numerical structure:

a) Goldstein condition [GO 1965]:

Let $\bar{\alpha} > 0$ be any number satisfying

$$0 < -\mu_1\bar{\alpha} \, \nabla_x F(x)^T d \leq F(x) - F(x+\bar{\alpha}d) \leq -\mu_2\bar{\alpha} \, \nabla_x F(x)^T d \qquad (8)$$

with arbitrary scalars μ_1, μ_2 so that $0 < \mu_1 \leq \frac{1}{2} \leq \mu_2 < 1$.

b) Armijo condition [AR 1966]:

Let $.5 > \mu > 0$ and $\bar{\alpha}$ be the largest number in the sequence $\{\sigma q^{-j}\}$ with $q > 1$, $1 \geq \sigma > 0$, $-\nabla_x F(x)^T d \geq \sigma \|d\| \|\nabla_x F(x)\|$, and

$$F(x) - F(x+\bar{\alpha}d) \geq -\mu\bar{\alpha} \, \nabla_x F(x)^T d \; . \qquad (9)$$

The corresponding convergence theory is found, e.g., in Ortega, Rheinboldt [OR 1970]. The practical importance of these or similar steplength algorithms is diminished by the possibility of getting a slow convergence rate. Nevertheless, conditions of the kind (8), (9), or related ones are often used as stopping criteria in one of the steplength procedures described later, cf. Bartholomew-Biggs [BI 1976], Murtagh, Sargent [MS 1969], or Gill [GI 1978].

1.2 Fibonacci and golden section methods

To explain the basic idea, it is assumed that the function

$$\varphi(\alpha) := F(x + \alpha d) \qquad (10)$$

is unimodal in the considered interval (a,b), i.e. φ has only one minimizer in (a,b). The Fibonacci search method determines two series of points s_k, t_k, k=1,1,... , to reduce the length of the uncertainty interval as much as possible. An uncertainty interval contains the minimizer $\bar{\alpha}$ of (7). If M denotes the maximum number of two-point searches and $\{f_k\}$ the Fibonacci sequence

$$f_{k+1} := f_k + f_{k-1} \; , \quad f_0 := f_1 := 1 \; , \quad k=1,2,\ldots \; ,$$

the points s_k, t_k are given by

$$s_{k+1} := (f_{M-1-k}/f_{M+1-k}) \cdot (\beta_k - \alpha_k) + \alpha_k$$
$$t_{k+1} := (f_{M-k}/f_{M+1-k}) \cdot (\beta_k - \alpha_k) + \alpha_k \qquad k=0,1,2,\ldots,M-1 \; .$$

Here, $\alpha_0 := a$, $\beta_0 := b$, and

$$\alpha_{k+1} := \alpha_k \; , \quad \beta_{k+1} := t_{k+1} \; , \quad \text{if } \varphi(s_{k+1}) < \varphi(t_{k+1}) \; ,$$

$$\alpha_{k+1} := s_{k+1} \; , \quad \beta_{k+1} := \beta_k \; , \quad \text{if } \varphi(s_{k+1}) \geq \varphi(t_{k+1}) \; .$$

It can be shown, cf. Ortega, Rheinboldt [OR 1970], that the quotient of the last uncertainty interval length (α_M, β_M) and the first one (a,b) is equal to $f_{M+1}-1$ so that the number of function evaluations is known a priori to bracket the minimum up to a predetermined accuracy.

Golden section search uses the strategy to decrease the length of the uncertainty interval by a constant factor τ leading to a simple algorithm of the following kind, cf. Kowalik and Osborne [KO 1968], for example:

Let $\alpha_0 := a$, $\beta_0 := b$, $s_0 := b - \tau(b - a)$, $t_0 := a + \tau(b - a)$.

If $\varphi(t_k) > \varphi(s_k)$, then

$$\beta_{k+1} := t_k, \; \alpha_{k+1} := \alpha_k \; ,$$
$$t_{k+1} := s_k$$
$$s_{k+1} := \alpha_k + (1 - \tau)(\beta_{k+1} - \alpha_k) \; .$$

If $\varphi(t_k) \leq \varphi(s_k)$, then

$$\alpha_{k+1} := s_k, \; \beta_{k+1} := \beta_k \; ,$$
$$s_{k+1} := t_k$$
$$t_{k+1} := \beta_k - (1 - \tau)(\beta_k - \alpha_{k+1})$$

for k=1,2,3,... . Here, τ denotes the golden section number

$$\tau := (\sqrt{5} - 1)/2 \approx .618 \ .$$

Search methods of the considered form have guaranteed convergence to the unidimensional minimizer $\bar{\alpha}$ at least for unimodal functions and an a priory bound for the number of function evaluations is known, but the major drawback is the fact that the rate of convergence is at best linear.

1.3 Interpolation methods

These methods use a quadratic or cubic fit for determining a suitable approximation of $\bar{\alpha}$, the minimizer of $\varphi(\alpha)$, cf. (10). Quadratic fit is preferred if only function evaluations are available. Given three arguments α_1, α_2, and α_3, the stationary point of the quadratic is given by

$$\alpha_4 := \frac{1}{2} \frac{\beta_{23}\varphi(\alpha_1) + \beta_{31}\varphi(\alpha_2) + \beta_{12}\varphi(\alpha_3)}{\gamma_{23}\varphi(\alpha_1) + \gamma_{31}\varphi(\alpha_2) + \gamma_{12}\varphi(\alpha_3)}$$

where $\beta_{ij} := \alpha_i^2 - \alpha_j^2$, $\gamma_{ij} := \alpha_i - \alpha_j$. If, on the other hand, gradient information is also available, it is possible to fit a cubic equation to two given points α_1 and α_2. The relative minimizer of the third order polynomial is

$$\alpha_3 := \alpha_2 - (\alpha_2 - \alpha_1) \frac{\varphi'(\alpha_2) + \beta_2 - \beta_1}{\varphi'(\alpha_2) - \varphi'(\alpha_1) + 2\beta_2}$$

with

$$\beta_1 := \varphi'(\alpha_1) + \varphi'(\alpha_2) - 3 \frac{\varphi(\alpha_1) - \varphi(\alpha_2)}{\alpha_1 - \alpha_2}$$

$$\beta_2 := (\beta_1^2 - \varphi'(\alpha_1)\varphi'(\alpha_2))^{1/2} \ .$$

For the choice of the points α_1, α_2, and α_3 for a quadratic fit or α_1, α_2 for a cubic fit, there exist various algorithms in the literature. They are based on different strategies to obtain the desired α_k so that

$$\varphi(\alpha_1) \geq \varphi(\alpha_2) \leq \varphi(\alpha_3)$$

or $\qquad\qquad \varphi(\alpha_1) \leq \varphi(\alpha_2) \ , \quad \varphi'(\alpha_1) < 0 \ ,$

respectively. In the first case, the reader is refered to Powell's search procedure, cf. Powell [PO 1964], Pierre [PI 1969], or Himmelblau [HI 1972]. A search technique using a cubic fit is derived by Davidon [DA 1959], see also Pierre [PI 1969]. It can be shown that methods of the considered type possess at least a super-linear local convergence, see Luenberger [LU 1965] or Brent [BT 1973].

1.4 Implementation of line-search algorithms

The implementation of a line-search algorithm is a crucial part of any optimization program, and the length of the corresponding subprogram could amount to several hundred statements. Only older programs like SUMT realize a golden section or similar method, whereas nearly all other programs under consideration perform line-searches with a quadratic fit, but with different strategies to determine the three interpolating points α_1, α_2, and α_3. To measure the efficiency of the steplength procedure, one should compare the average number of function evaluations with the average number of gradient evaluations. The quotient of both numbers could be more than 20 for golden section routines (SUMT) and approximately 1 for interpolating methods (SALQDR, VF02AD, OPRQP, XROP) indicating that it is possible to derive search directions with steplength one in most iterations.

Programs realizing line-search algorithms are part of the modular system of software packages to solve constrained problems of the kind (1), cf. Chapter III. Freely available listings are published by Lootsma [LO 1978], Pierre and Lowe [PL 1975], or Kuester and Mize [KM 1973]. Some line-search methods are described in Himmel-blau [HI 1972] together with numerical comparisons. An efficient and sophisticated algorithm is presented by Gill [GI 1978]. Powell [PO 1978a] proposed a quite simple and short line-search procedure which is nevertheless very efficient to implement in a program

solving problem (1). However, one has to be aware that it is not
always possible to use a steplength procedure independently of
the underlying algorithm. In particular penalty terms in the function
F could lead to a pathological curvature of φ requiring safeguarded
interpolation schemes.

2. Quadratic programming

The problem under consideration is to minimize a quadratic
objective function subject to linear equality and inequality constraints:

$$\min \ \frac{1}{2} x^T C x + d^T x$$
$$x \in \mathbb{R}^n: \begin{array}{l} A_1 x \leq b_1 \\ A_2 x = b_2 \end{array} \tag{11}$$

C is an (n,n) matrix, $d \in \mathbb{R}^n$, $b_1 \in \mathbb{R}^{m_1}$, $b_2 \in \mathbb{R}^{m_2}$, and A_1, A_2 are
(m_1,n) and (m_2,n) matrices, respectively. Quadratic programming
algorithms are considered in the context of this report since they
appear as subproblems in solution methods for the general constrained
optimization problem (1).

2.1 Equality constrained problems

When we assume that $m_1 = 0$ and C is positive definite, then the
objective function in (11) is strictly convex and the unique solution
x* of the equality constrained quadratic program

$$\min \ \frac{1}{2} x^T C x + d^T x$$
$$x \in \mathbb{R}^n: A x = b \tag{12}$$

is obtained by

$$x^* = x_0 - (C^{-1} - C^{-1} A^T (A C^{-1} A^T)^{-1} A C^{-1}) g \tag{13}$$

where $g = Cx_0 + d$ and x_0 is any feasible point, see, e.g., Powell [PO 1974]. The optimality of x* is easily derived by differentiating the Lagrange function of problem (12) with respect to the variable x and the optimal multipliers $u* = (AC^{-1}A^T)^{-1}AC^{-1}g$. The calculations in (13) are carried out by any decomposition technique. In other words, the solution of the equality constrained quadratic program is obtained by some matrix manipulations.

2.2 Projection methods

Starting from a feasible iterate x_0, methods of this type generate a sequence of equality constrained subproblems of the kind (12) which are obtained by projecting the minimum of the objective function $\frac{1}{2} x^T Cx + d^T x$ on the linear manifold of the active constraints. A constraint $a_i^T x \leq b_i$ is called active with respect to x_0, if $a_i^T x_0 = b_i$. If an inactive constraint is violated by the projected minimum, say $\bar{x}$, then the constraint is added to the set of active constraints and the last feasible point in the direction pointing to $\bar{x}$ is the new iterate. If, on the other hand, $\bar{x}$ is feasible, but violates the optimality conditions, a constraint is dropped from the set of active constraints and the whole procedure is repeated. The adding and dropping steps are performed by special update techniques.

Projection methods are fitted to solve problems with an indefinite matrix C as well, cf. Fletcher [FL 1971] or Gill and Murray [GM 1977]. If a Cholesky decomposition of C in the form LL^T is known, the quadratic program (11) is identical with a linear least squares problem. Projection methods for solving linear least squares problems

$$\min \quad \|Ex - f\|$$
$$x \in \mathbb{R}^n: \quad \begin{aligned} A_1 x &\leq b_1 \\ A_2 x &= b_2 \end{aligned}$$

are presented in Lawson, Hanson [LH 1974] or Schittkowski, Stoer [SS 1979].

2.3 Linear programming based methods

To explain the basic idea, it is more convenient to start from the following formulation of the quadratic program:

$$\min \frac{1}{2} x^T C x + d^T x$$

$$x \in \mathbb{R}^n: \begin{array}{l} Ax \leq b \\ x \geq 0 \end{array}$$

For most algorithms, it is assumed that the matrix C is positive semi-definite. A solution method for general quadratic optimization problems is given by Keller [KE 1973]. The Kuhn-Tucker condition shows that x* is an optimal solution of (15) if and only if there are vectors y*, u*, v* satisfying

$$\begin{pmatrix} 0 & : & -A \\ A^T & : & C \end{pmatrix} \begin{pmatrix} y* \\ x* \end{pmatrix} + \begin{pmatrix} b \\ d \end{pmatrix} = \begin{pmatrix} u* \\ v* \end{pmatrix} \tag{16}$$

with $u* \geq 0$, $v* \geq 0$, $x* \geq 0$, $y* \geq 0$, and $y*^T u* + x*^T v* = 0$. Linear programming based methods try to solve the equivalent problem (16), sometimes denoted as the complementary problem. The algorithms generate a sequence of feasible basis solutions by simplex transformations until the optimality criteria are fulfilled. One of the first methods was given by Wolfe [WO 1959], cf. also Koo [KO 1977], which was modified by Dantzig [DZ 1963]. A successful method was developed by Beale [BL 1967]. Some comments about differences and common features of linear programming based methods and projection methods are given by Fletcher [FL 1971].

2.4 Implementation of quadratic programming algorithms

The projection method of Fletcher [FL 1971] for general quadratic optimization problems (11) was implemented for the Harwell Subroutine Library. The similar method of Gill and Murray [GM 1977] is part of the Optimization Software Library of the National Physical Laboratory. ALGOL procedures for solving linear least squares problems of

the kind (14) are published by Schittkowski, Zimmermann [SZ 1977]
and FORTRAN programs are contained in Lawson, Hanson [LH 1974]. A
program realizing Wolfe's [WO 1959] method for quadratic problems
in the form (15) is found in Kuester, Mize [KM 1973], and Land and
Powell [LP 1973] present a program based on Beale's [BL 1967]
method.

3. Unconstrained optimization

It will be shown in the subsequent sections that many optimization methods require the solution of an unconstrained nonlinear
programming subproblem of the kind

$$\min_{x \in \mathbb{R}^n} F(x) \tag{17}$$

with a continuously differentiable function $F: \mathbb{R}^n \to \mathbb{R}$. The most
efficient algorithms for solving (17) generate a sequence of iterates
$\{x_k\}$ so that

$$x_{k+1} := x_k + \alpha_k p_k \; . \tag{18}$$

The starting point x_0 is chosen arbitrarily, the steplength α_k is
calculated by one of the line-search algorithms of Section 1, and
the search direction p_k is obtained by a strategy described in the
following subsections. These techniques can be found in any standard book on nonlinear programming, cf. for example Himmelblau
[HI 1972] or Luenberger [LU 1965]. More timely surveys of some
unconstrained optimization techniques are given by Dennis and Moré
[DM 1977] and Gill [GI 1978].

3.1 Conjugate direction algorithms

An unconstrained optimization method is called a conjugate direction algorithm, if the search directions p_k are conjugate when minimizing a quadratic function of the kind

$$F(x) := \frac{1}{2} x^T A x + b^T x ,$$ (19)

i.e. if

$$p_j^T A p_i = 0$$ (20)

for $i=0,1,\ldots,n-1$, $j=0,1,\ldots,n-1$, $i \neq j$. It can be shown that a conjugate direction algorithm for minimizing a strictly convex quadratic function with exact line-search, cf. (7), is finite and attains the minimum of F in at most n iterations.

The most popular method for constructing conjugate directions is that of Fletcher and Reeves [FR 1964]:

$$p_0 := -\nabla_x F(x_0) ,$$

$$p_k := -\nabla_x F(x_k) + \gamma_k p_{k-1} ,$$

$$\gamma_k := \|\nabla_x F(x_k)\|^2 / \|\nabla_x F(x_{k-1})\|^2 ,$$

for $k=1,2,\ldots$. There exist a lot of related formulae to obtain conjugate directions, cf. Gill [GI 1977], for example, and the method can be expanded to solve optimization problems in Hilbert spaces, cf. Daniel [DL 1971].

Numerical experience, however, shows that conjugate direction methods are sensitive with respect to the choice of the steplength. If α_k is only a poor approximation of the exact minimizer, these methods become inefficient and are inferior to quasi-Newton algorithms which are described subsequently. Nevertheless they possess the advantage that no matrices are required to compute the search direction p_k so that conjugate direction methods can be implemented to solve large problems, cf. Peschon, Peterson [PP 1975] or Lasdon, Waren [LW 1977].

3.2 Quasi-Newton algorithms

Quasi-Newton or variable metric methods determine the search direction p_k by multiplying the gradient with the inverse of a matrix B_k, i.e.

$$p_k := - B_k^{-1} \nabla_x F(x_k) \ ,$$

where B_k denotes the Hessian matrix of F at x_k, if second partial derivatives are available, or otherwise any suitable approximation. In the last case, B_k is updated during every iteration step by a rank-two correction of the form

$$B_{k+1} := B_k + u_k v_k^T + \bar{u}_k \bar{v}_k^T$$

under the additional requirements that B_k remains symmetric, positive definite, and that the quasi-Newton condition

$$B_{k+1} s_k = q_k$$

with $s_k := x_{k+1} - x_k$, $q_k := \nabla_x F(x_{k+1}) - \nabla_x F(x_k)$ is satisfied. Among all possibilities of fulfilling these conditions, extensive numerical experiments of recent years indicate that the BFGS-formula, cf. Broyden [BY 1970], Fletcher [FL 1970], Goldfarb [GF 1970], Shanno [SO 1970], given by

$$B_{k+1} := B_k + \frac{q_k q_k^T}{q_k^T s_k} - \frac{B_k s_k s_k^T B_k}{s_k^T B_k s_k} \tag{21}$$

seems to be the most effective one. As a starting matrix B_0 one could define any symmetric, positive definite matrix, for example I. For the numerical realization, B_k is often factorized into

$$B_k := L_k L_k^T \tag{22}$$

with a lower triangular matrix L_k. Instead of calculating the expression (21), the matrix L_k can be updated with about the same amount of numerical operations, see Gill, Murray, and Saunders [GS 1975]. If F is a strictly convex quadratic function and if exact line-search for determining α_k is used, then the search directions p_k are conjugate and the algorithm stops after at most n iterations.

If one wants to avoid the factorization (22), it is possible to derive analogous updating schemes for the matrices $H_k := B_k^{-1}$. A well-known process is the DFP-formula, cf. Davidon [DA 1959], Fletcher and Powell [FP 1963], given by

$$H_{k+1} := H_k + \frac{s_k s_k^T}{q_k^T s_k} - \frac{H_k q_k q_k^T H_k}{q_k^T H_k q_k} \quad .$$

3.3 Implementation of unconstrained nonlinear programming methods

Every qualified optimization library contains programs for solving unconstrained optimization problems. In particular, the library of the National Physical Laboratory contains an extensive series of FORTRAN programs for solving problems of the kind (17) with different assumptions about the differentiability of F or with special types of objective functions. Most programs realize the BFGS-formula or related optimally-conditioned updating formulae which try to minimize the condition number of B_k, cf. Davidon and Nazareth [DN 1977]. Some further programs are contained in the books of Himmelblau [HI 1972] and Kuester, Mize [KM 1973].

4. Penalty methods

Penalty methods belong to the first attempts to solve constrained optimization problems of the form (1) satisfactorily. The fundamental idea is to construct a sequence of unconstrained problems and to solve them by one of the techniques described in the previous section so that the minimizers of the unconstrained problems converge to a solution of the constrained one. For this, so called penalty terms are added to the objective function which 'penalize' f whenever the feasible region is left. A factor r_k controls the degree of penalizing f. Proceeding from a sequence $\{r_k\}$ with $r_k \to \infty$ for $k \to \infty$, penalty functions can be defined for example by

$$P(x,r_k) := f(x) + r_k \sum_{j=1}^{m_e} g_j(x)^2 + r_k \sum_{j=m_e+1}^{m} (\min(0,g_j(x))^2 \qquad (23)$$

or

$$P(x,r_k) := f(x) + r_k \sum_{j=1}^{m_e} g_j(x)^2 - \frac{1}{r_k} \sum_{j=m_e+1}^{m} \ln(g_j(x)) \qquad (24)$$

The function (23) is called an exterior penalty function, since, in general, the minimizer is an exterior point of the feasible region, and (24) gives an interior penalty function, since the iterates are prevented from reaching the inequality bounds. Combination of (23) and (24) gives the so called mixed interior-exterior penalty function

$$P(x,r_k) := f(x) + r_k \sum_{j=1}^{m_e} g_j(x)^2 + r_k \sum_{j\in J_1} (\min(0,g_j(x))^2$$

$$- \frac{1}{r_k} \sum_{j\in J_2} \ln(g_j(x))$$

where J_1 and J_2 are defined by

$$J_1 := \{j: g_j(x_0) \leq 0, \ m_e < j \leq m\}$$
$$J_2 := \{j: g_j(x_0) > 0, \ m_e < j \leq m\} \ .$$

Here, x_0 denotes the starting point of the algorithm. Minimizing (23), (24), or any related expression yields an approximate solution of (1). Under some additional assumptions, convergence of the algorithm to the solution of the constrained problem can be shown, cf. Fiacco and McCormick [FM 1968].

Penalty methods can be implemented easily in a program library provided that an unconstrained optimization algorithm is available. From the present point of view, penalty methods are inferior to the subsequently developed multiplier, quadratic approximation, or generalized reduced gradient methods. The main disadvantage is the increasing condition number of the Hessian matrix of the penalty function generating more and more ill-conditioned unconstrained sub-problems, cf. Murray [MU 1967] or Lootsma [LO 1971].

5. Multiplier methods

Multiplier methods, sometimes called penalty-multiplier or augmented Lagrangian methods, try to avoid the disadvantage of penalty algorithms that the parameters r_k have to tend to infinity in order to achieve convergence. They were first developed to solve equality constrained problems, i.e. $m_e = m$, by minimizing a sequence of modified penalty functions. Powell [PO 1969] defined them by

$$\Phi(x,v,s) := f(x) + \frac{1}{2} \sum_{j=1}^{m} s_j (g_j(x) - v_j)^2 \tag{25}$$

where $x \in \mathbb{R}^n$, $v = (v_1,\dots,v_m)^T \in \mathbb{R}^m$, $s = (s_1,\dots,s_m)^T \in \mathbb{R}^m$. By choosing sufficiently large s_j-values and a suitable updating scheme for v after each unconstrained minimization process, it is hoped that $v_j s_j$ approximates u_j^*, the j-th optimal Lagrange multiplier of (1). Hestenes [HE 1969] suggested using the following augmented Lagrange function for solving (1):

$$\Psi(x,u,s) := f(x) - \sum_{j=1}^{m} u_j g_j(x) + \frac{1}{2} \sum_{j=1}^{m} s_j g_j(x)^2 \tag{26}$$

with $x \in \mathbb{R}^n$, $u = (u_1,\dots,u_m)^T \in \mathbb{R}^m$, $s = (s_1,\dots,s_m)^T \in \mathbb{R}^m$. Letting $s_j v_j = u_j$, (25) and (26) are correlated by

$$\Phi(x,v,s) = \Psi(x,u,s) + \frac{1}{2} \sum_{j=1}^{m} s_j v_j^2 \ , \tag{27}$$

so that both functions possess the same minimizers with respect to x.

If inequality restrictions appear, i.e. if $m_e < m$, (25) and (26) have to be modified in an analogous way. Fletcher [FL 1975] adapted Powell's function (25) to inequality constraints by defining

$$\Phi(x,v,s) := f(x) + \frac{1}{2} \sum_{j=1}^{m_e} s_j (g_j(x) - v_j)^2$$
$$+ \frac{1}{2} \sum_{j=m_e+1}^{m} s_j (g_j(x) - v_j)_-^2 \tag{28}$$

where $a_- := \min(0,a)$ for $a \in \mathbb{R}$. A suitable modification of the Hestenes function (26) is given by Rockafellar [RO 1974]:

$$\Psi(x,u,s) := f(x) - \sum_{j=1}^{m_e} (u_j g_j(x) - \tfrac{1}{2} s_j g_j(x)^2) \tag{29}$$

$$- \sum_{j=m_e+1}^{m} \left\{ \begin{array}{ll} (u_j g_j(x) - \tfrac{1}{2} s_j g_j(x)^2) \ , & \text{if } g_j(x) \le u_j/s_j \\ \tfrac{1}{2} u_j^2/s_j & , \quad \text{otherwise} \end{array} \right.$$

Again, (27) remains true by letting $u_j = s_j v_j$, $j=1,\dots,m$.

A multiplier method starts with sufficiently large penalty parameters s_j, $j=1,\dots,m$, and some initial guesses for v or u, respectively. The iterative process consists of solving the unconstrained minimization problems

$$\min \ \Phi(x,v,s)$$
$$x \in \mathbb{R}^n$$

or

$$\min \ \Psi(x,u,s)$$
$$x \in \mathbb{R}^n$$

leading to a new approximation $\bar{x}$. The updating schemes for v and u are given by simple rules of the kind

$$\bar{v}_j := v_j - g_j(\bar{x}) \ , \quad j=1,\dots,m_e$$
$$\bar{v}_j := v_j - \min(g_j(\bar{x}),v_j) \ , \quad j=m_e+1,\dots,m \ ,$$

or

$$\bar{u}_j := u_j - s_j g_j(\bar{x}) \ , \quad j=1,\dots,m_e$$
$$\bar{u}_j := u_j - \min(s_j g_j(\bar{x}),u_j) \ , \quad j=m_e+1,\dots,m \ ,$$

respectively. The penalty terms s_j are only increased if the convergence is not as rapid as required. It can be shown that a solution x^* of (1) and the corresponding optimal Lagrange multipliers u^* are approximated with a local linear rate of convergence provided that s_j is sufficiently large, $j=1,\dots,m$.

6. Quadratic approximation methods

A quadratic approximation method requires the solution of a quadratic subproblem in each iteration step to determine a search direction. Subsequently, a penalty function is minimized along this search direction leading to a new iterate. The objective functions in the subproblems consist of a quadratic approximation of either the objective function f or the Lagrangian L, but we can observe different strategies to define the linear restrictions.

6.1 The method of Murray and Biggs

The algorithm proceeds from an exterior penalty function of the kind (23) written in the form

$$P(x,r_k) := f(x) + r_k \, g(x)^T g(x)$$
$$= f(x) + r_k \sum_{j \in J_x} g_j(x)^2 \tag{30}$$

where J_x denotes an index set of so called active constraints, $g(x)$ the vector of all active constraints, and $\{r_k\}$ is a sequence of increasing penalty parameters. The fundamental idea is to compute approximate minima of (30) rather than precise ones as done by the classical penalty methods of Section 4. Given any iterate x, differentiation of (30) leads to

$$\nabla_x P(x,r_k) = \nabla_x f(x) + 2r_k \, D_x g(x)^T g(x) . \tag{31}$$

One now requires a correction p of x such that x + p could be considered as an approximation of the unconstrained minimizer of (30). For this the truncated Taylor series of (31)

$$\nabla_x P(x+p,r_k) \doteq \nabla_x f(x) + \nabla_x^2 f(x)p + 2r_k \, D_x g(x)^T g(x)$$
$$+ 2r_k \, D_x g(x)^T D_x g(x)p + 2r_k \sum_{j \in J_x} g_j(x)\nabla_x^2 g_j(x)p \tag{32}$$

should vanish.

Neglecting the term $\nabla_x^2 f(x)p$, Murray [MU 1969a, MU 1969b] shows that a first order estimate of (32) can be characterized by

$$D_x g(x)p \geq \left(\frac{r_{k-1}}{r_k} - 1 \right) g(x) \quad . \tag{33}$$

These inequalities define the linear restrictions of the quadratic subproblem in Murray's method. The objective function is given by

$$\frac{1}{2} p^T B p + p^T \nabla_x f(x)$$

where B denotes a suitable approximation of the Hessian of $f(x)$ which is updated in every iteration by a quasi-Newton algorithm. The integer set J_x contains all indices belonging to violated constraints and there are suitable updating schemes for the penalty parameters r_k, see Murray [MU 1969b] or Biggs [BI 1971].

Biggs [BI 1972, BI 1976] modified Murray's method in the sense that (32) is set to zero leading to the equation

$$(G(x) + 2r_k \, D_x g(x)^T D_x g(x))p = - \nabla_x f(x) - 2r_k \, D_x g(x)^T g(x)$$

with

$$G(x) := \nabla_x^2 f(x) + 2r_k \sum_{j \in J_x} g_j(x) \nabla_x^2 g_j(x) \quad .$$

As a consequence, we have

$$D_x g(x)p = - \frac{1}{2r_k} v - g(x) \quad ,$$

$$\left(\frac{1}{2r_k} I + D_x g(x) G(x)^{-1} D_x g(x)^T \right) v = D_x g(x) G(x)^{-1} \nabla_x f(x) - g(x) \tag{34}$$

To realize (34) numerically, $G(x)^{-1}$ is replaced by any approximation H for which quasi-Newton updating schemes are available. The vector v tends to the optimal Lagrange multipliers of the original problem (1). Using (34) as linear equality restrictions and a quadratic approximation of the penalty function (30), the method requires the solution of a quadratic subprogram

$$\min \quad \frac{1}{2} p^T H^{-1} p + \nabla_x f(x)p$$

$$D_x g(x)p = - \frac{1}{2r_k} v - g(x)$$

which is easily calculated by

$$p = H(D_x g(x)^T v - \nabla_x f(x)) \quad .$$

The index set J_x now contains the indices of the violated constraints and, in addition, of all inequality constraints violated at the previous iterate with $v_j > 0$. Biggs [BI 1976] gives some rules for calculating the penalty parameter r_k, and in [BI 1978] some convergence properties are collected.

6.2 The method of Wilson, Han, and Powell

The algorithm defines a quadratic subproblem in each iteration step to minimize an approximation of the Lagrangian subject to the linearized constraints of (1). Given a current iterate x, a correction d is obtained by solving

$$\begin{aligned}
\min \quad & \tfrac{1}{2} d^T B d + \nabla_x f(x)^T d \\
d \in \mathbb{R}^n: \quad & g_j(x) + \nabla_x g_j(x)^T d = 0 , \quad j=1,\dots,m_e , \\
& g_j(x) + \nabla_x g_j(x)^T d \geq 0 , \quad j=m_e+1,\dots,m .
\end{aligned} \tag{35}$$

The method was first studied by Wilson [WI 1963] with $B := \nabla_x^2 L(x,u)$. The Lagrange multipliers of (35) are used as guesses for the multipliers of the original problem for defining B. Han [HN 1976] replaced the Hessian of L by quasi-Newton update procedures and proved the local superlinear convergence. Powell [PO 1978a] implemented the method in an efficient way. In particular, he updated B by the BFGS-formula (21) and introduced a line-search to determine the new iterate in the form

$$\bar{x} := x + \alpha d$$

where d is the solution of (35) and α the steplength obtained by a quadratic interpolation of a non-differentiable penalty function. Some convergence properties of this method are given in Powell [PO 1978b].

7. Generalized reduced gradient methods

By introducing non-positive slack variables, the original non-linear programming problem (1) is equivalent to another type of optimization problem with nonlinear equality constraints only and bounds on the variables:

$$\min \quad F(x)$$
$$x \in \mathbb{R}^N: \quad \begin{aligned} & G_j(x) = 0 \ , \quad j=1,\ldots,M \\ & l \leq x \leq u \end{aligned} \qquad (36)$$

Some of the bounds l_i, u_i, $i=1,\ldots,N$, could be $-\infty$ or $+\infty$, respectively.

To explain the fundamental structure of generalized reduced gradient methods, one should consider them as generalizations of linear programming algorithms, for example the simplex method or its variants. Indeed, the first programs were written to solve linearly constrained problems, cf. Wolfe [WO 1963, WO 1967] or Faure, Huard [FH 1965]. Let x be any feasible point of problem (36). This variable can be partitioned into a so-called basis variable $x^b \in \mathbb{R}^M$ and a non-basis variable $x^{nb} \in \mathbb{R}^{N-M}$. For simplicity, they are ordered so that

$$x = \begin{pmatrix} x^b \\ x^{nb} \end{pmatrix} .$$

Analogously, the upper and lower bounds are partitioned into

$$l = \begin{pmatrix} l^b \\ l^{nb} \end{pmatrix} , \quad u = \begin{pmatrix} u^b \\ u^{nb} \end{pmatrix} ,$$

and the Jacobian matrix of $G := (G_1,\ldots,G_M)^T$ into

$$D_x G(x) = (D_{x^b} G(x) : D_{x^{nb}} G(x)) .$$

If the basis variables are chosen so that the square matrix $D_{x^b} G(x)$ is nonsingular, then the equations

$$G_j(x^b, x^{nb}) = 0 \ , \quad j=1,\ldots,M , \qquad (37)$$

could be solved at least conceptually to obtain x^b for any given non-basis variables x^{nb}. So (36) is transformed into a simpler reduced problem

$$\min \quad F*(x^{nb})$$
$$x^{nb} \in \mathbb{R}^{N-M}: \quad l^{nb} \leq x^{nb} \leq u^{nb} \tag{38}$$

where $F*$ is defined by

$$F*(x^{nb}) := F(x^b(x^{nb}), x^{nb})$$

and $x^b(x^{nb})$ is a solution of (37) for a given independent variable x^{nb}. The gradient of $F*$ is the so-called reduced gradient of F given by

$$z^{nb} := \nabla_{x^{nb}} F*(x^{nb}) = \nabla_{x^{nb}} F(x) - D_{x^{nb}} G(x)^T (D_{x^b} G(x)^T)^{-1} \nabla_{x^b} F(x) \tag{39}$$

for some $x = \begin{pmatrix} x^b \\ x^{nb} \end{pmatrix} \in \mathbb{R}^N$.

A generalized reduced gradient method proceeds from a feasible iterate x partitioned into

$$x = \begin{pmatrix} x^b \\ x^{nb} \end{pmatrix}.$$

With the goal to solve (38), the reduced gradient (39) or one of the well-known modifications like conjugate gradient or quasi-Newton directions is computed, leading to a vector $\bar{d} \in \mathbb{R}^{N-M}$. But this direction has to be projected on the parallelotope defined in (38) leading to the search direction $d = (d_1, \ldots, d_{N-M})^T$ with

$$d_i := \begin{cases} 0, & \text{if } x_i^{nb} = l_i^{nb} \text{ and } z_i^{nb} > 0, \\ 0, & \text{if } x_i^{nb} = u_i^{nb} \text{ and } z_i^{nb} < 0, \\ \bar{d}_i, & \text{otherwise,} \end{cases}$$

$i=1, \ldots, N-M$. Then a one-dimensional search procedure has to be applied to get an approximate solution of

$$\min \quad F*(x^{nb} + \alpha d)$$
$$\alpha \geq 0$$

When realizing the line-search by one of the algorithms described in Section 1, the systems

$$x^b : \quad G_j(x^b, x^{nb} + \alpha d) = 0 \ , \quad j=1,\ldots,M,$$

have to be solved for some values of α. This is a system of nonlinear equations with M unknown variables x_i^b, $i=1,\ldots,M$, and can be
solved by Newton's method or one of its variants to reduce the
number of gradient evaluations. Assuming now that this method con-
verges, one has to be aware of the possibility that some basis variables may violate their bounds. In other words, the line-search
process can terminate with the request to adjust the basis variables and to restart the solution with a new reduced problem.

Many details of numerical realization of line-search algorithms
for generalized reduced gradient methods differ in existing optimization programs and the reader is referred to the concepts of Lasdon,
Waren [LW 1978] and Lasdon, Waren, Jain, Ratner [LJ 1978] or that
of Abadie [AB 1978] and Abadie, Carpentier [AC 1969]. Generalized
reduced gradient methods solve not only small, highly nonlinear
optimization problems, but are also adapted to solve large, sparse
problems with thousands of variables and restrictions, cf. Peschon,
Peterson [PP 1975], Murtagh, Saunders [MD 1976], or Lasdon, Waren
[LW 1978].

8. The method of Robinson

Robinson [RB 1972] developed an algorithm for solving nonlinear
optimization problems which does not fall into one of the categories
described so far. The fundamental idea is the construction of a
sequence of linearly constrained subproblems in which the Lagrange
function $L(x,u)$ is minimized and the constraints linearized as in
the method of Wilson, Han, and Powell, cf. (35):

$$\min \quad L(x,u^k)$$
$$x \in \mathbb{R}^n: \quad
\begin{array}{l}
g_j(x^k) + \nabla_x g_j(x^k)^T(x - x^k) = 0 , \quad j=1,\dots,m_e \\[2mm]
g_j(x^k) + \nabla_x g_j(x^k)^T(x - x^k) \geq 0 , \quad j=m_e+1,\dots,m
\end{array} \qquad (40)$$

x^k is the current iterate and $u^k := (u_1^k,\dots,u_m^k)^T$ an estimate for the
Lagrange multipliers. It is easy to see that (40) is equivalent to

$$\min \quad f(x) - \sum_{j=1}^{m} u_j^k(g_j(x)-g_j(x^k)-\nabla_x g_j(x^k)^T(x-x^k))$$
$$x \in \mathbb{R}^n: \quad
\begin{array}{l}
g_j(x^k) + \nabla_x g_j(x^k)^T(x - x^k) = 0 , \quad j=1,\dots,m_e \\[2mm]
g_j(x^k) + \nabla_x g_j(x^k)^T(x - x^k) \geq 0 , \quad j=m_e+1,\dots,m .
\end{array} \qquad (41)$$

The reason for using (41) instead of (40) is that if x^k and u^k
are replaced by the optimal $x*$ and $u*$, then $x*$ solves (41) and
$u*$ is an optimal Lagrange multiplier vector of (41). Therefore, a
solution of (41) gives the new iterate x^{k+1} and, in addition, the
corresponding Lagrange multipliers a new suitable guess u^{k+1}. To
implement this method numerically, a subroutine for minimizing
nonlinear objective functions subject to linear constraints has
to be available.

The subproblem (41) is closely related to the quadratic subproblem
(35) in the method of Wilson, Han, and Powell. The only difference
is that the Lagrangian in (41) has been replaced by a quadratic
approximation of the form

$$\frac{1}{2}(x - x^k)^T B^k(x - x^k) + \nabla_x L(x^k,u^k)^T(x - x^k)$$

to get the objective function in (35) by setting $d = x - x^k$.

OPTIMIZATION PROGRAMS

In this chapter, we describe the submitted optimization programs. In particular, we summarize technical details concerning program organization and present additional information such as author, source, algorithm, etc.. All programs are written in FORTRAN and were obtained from the authors in their original form.

1. Program organization

For most practical applications, an optimization program has to satisfy special technical assumptions. A user might be interested in a program in which the problem functions are provided in a certain way, with embedded numerical differentiation, with variable array dimensions, or similar requirements. In other words, the first step for selecting an optimization code consists of investigating the program organization with regard to the possible applications of the code. A review of all these technical details is presented in Table 1, but first we have to explain the notation:

LEN : Contains the approximate number of FORTRAN statements.

PRE : Gives information about the original precision of the programs.
    SP - Single precision.
    DP - Double precision.
    SDP - Single and double precision versions available on request.

DAT : Describes the provision of problem data such as dimension, number of constraints, parameters, etc..
    D   - All problem data are delivered in the driving program.

C   -   The problem data are read in on card reader unit number.

S   -   The code is implemented in the form of a main program and problem data are to be defined in a user-written subroutine.

FUN : Declares how the problem functions and their derivatives have to be provided. This information is especially important when a user wishes to deliver data or to calculate functions or gradients within a block. The provision of the problem functions is described by special formats of the kind

$$E_1 \ \& \ E_2 \ \& \ \dots \ \& \ E_s \ ,$$

showing that $E_1, \dots, E_s$ have to be evaluated altogether in a block. Each E_i stands for one of the following symbols:

F   -   Objective function.

G   -   All constraint functions.

GJ   -   One or a part of the constraint functions.

DF   -   Gradient of the objective function.

DG   -   Gradients of all constraint functions.

DGJ -   Gradient of one or a part of the constraint functions.

To give an example, consider the expressions

$$F \ \& \ G, \ DF \ \& \ DG$$

indicating that objective function and all restrictions are evaluated together, and, separately, the gradients of the objective function and all restrictions.

BOU : Provision of upper and lower bounds of the variables.

Y   -   Bounds are handled separately.

N   -   Bounds have to be formulated as inequality constraints.

DIM : Information on dimensioning of arrays.

FD   -   The program uses fixed dimensions for all arrays.

VD   -   The dimensions are completely variable.

MD   -   The program uses fixed dimensions to a certain degree, but the dimensions can easily be adapted to solve higher dimensional problems.

PAR : Number of input parameters such as stopping tolerances, maximum iteration number, initial penalty factor, etc., which are set by the user and which do not possess default values.

All data predetermined by the problem such as dimension, number
of constraints, print controls, etc., are not counted.

DIF : Information about the existence of enclosed numerical differen-
tiation subroutines.
  A   -   Analytical first derivatives are required, no numerical
          differentiation subroutine included.
  N1 -   Numerical approximations of the gradients are enclosed.
  N2 -   Analytical second partial derivatives can be provided
          or the embedded numerical approximation used.

MET : Declares if the user is allowed to choose between several
modifications of the mathematical algorithm.
  Y   -   Yes.
  N   -   No.

STO : Declares if several stopping rules could be used.
  Y   -   Yes.
  N   -   No.

FLA : Determines if one has the possibility to terminate the execution
of a problem immediately during the call of one of the user
provided subroutines and to return to the driving program.
  Y   -   Yes.
  N   -   No.

LIS : Declares if the FORTRAN listings are published in a freely
available form.
  Y   -   Yes.
  N   -   No.

ALG : Denotes the underlying mathematical method:
  QU -   Quadratic approximation method.
  GRG -   Generalized reduced gradient method.
  MU -   Multiplier method.
  PE -   Penalty method.
  RO -   Robinson's method

PAG : The last column shows the page number of the subsequent section,
where the program is described in more detail.

Code	LEN	PRE	DAT	FUN	BOU	DIM	PAR	DIF	MET	STO	FLA	LIS	ALG	PAG
OPRQP/XROP	500	SP	D	F&G,DF&DG	N	FD	4	A	N	N	N	N	QU	32
VFO2AD	1000	DP	D	F&G&DF&DG	N	VD	2	A	N	N	Y	N	QU	33
GRGA	4000	SP	D	F,G,DF,DG	Y	FD	0	N1	Y	N	N	N	GRG	34
OPT	1200	SP	D	F,G,DG	Y	VD	5	N1	N	N	N	N	GRG	35
GRG2(1/2)	4400	DP	C	F&G,DF&DG	Y	VD	0	N1	Y	N	N	N	GRG	36
VFO1A	950	SDP	D	F&G&DF&DG	N	MD	4	A	N	N	N	N	MU	37
LPNLP	1200	SP	C	F&G,DF&DG	Y	VD	12	A	Y	N	N	Y	MU	38
SALQDR/SALQDF/ /SALMNF	27000*	SDP	D	F,DF,F&DF, G,G&DG	Y	VD	5	N2	Y	N	Y	N	MU	39
CONMIN	850	SP	D	F&G,DF&DG	N	FD	8	N1	N	N	N	Y	MU	40
BIAS(1/2)	850	SP	D	F,G,DF&DG	Y	VD	5	N1	N	N	N	N	MU	41
FUNMIN	1200	DP	D	F&G,DF&DG	N	VD	3	A	Y	N	N	N	MU	42
GAPFPR	1000	SP	C	F&G&DF&DG	Y	FD	7	A	N	N	N	N	MU	43
GAPFQL	450	SP	S	F,GJ,DF,DGJ	Y	FD	0	A	N	Y	N	N	MU	44
ACDPAC	3900	DP	C	F,GJ,DF,DGJ	Y	VD	0	A	N	N	N	N	MU	45
FMIN(1/2/3)	1300	SP	D	F,GJ,DF,DGJ	Y	VD	6	N2	Y	Y	N	Y+	PE	46
NLP	7000*	SP	D	F&G,F&G&DF&DG	N	FD	5	N1	Y	Y	Y	N	PE	47
SUMT	1500	SP	C	F,GJ,DF,DGJ	Y	FD	7	N2	Y	Y	Y	Y	PE	48
DFP	1000	DP	D	F,F&G&DF&DG	N	FD	4	A	Y	N	N	N	PE	49
FCDPAK	3200	DP	C	F,GJ,DF,DGJ	Y	VD	0	A	Y	N	N	N	RO	50

Table 1: Program organization.

* The programs are part of a major optimization package and cannot be obtained separately.

+ The equivalent ALGOL 60 procedures are listed in Lootsma [LO 1978].

2. Description of the programs

In this section, we give further individual information about the
programs such as author, source, references, etc.. In particular, we
briefly list their most important performance properties - Chapter V
gives further details. Performance criteria are efficiency, reliability, global convergence, performance for solving degenerate, ill-conditioned, and indefinite problems, sensitivity to slight variations
of the problem or to the position of the starting point, and ease of
use. The programs are described using the following scheme:

NAME :   Name of the program and its tested versions.

AUTHOR :   Name of the author(s).

SOURCE :   Address, reference, or institution from which the
                program was obtained.

ALGORITHM :   The mathematical background is characterized and the
                reader is referred to the corresponding section of
                Chapter II.

VERSIONS :   If a program is executed in different versions as
                indicated in the NAME-row, we identify and describe
                these versions.

REFERENCE :   Some references are given which describe the usage
                of the program and the underlying mathematical method.

PERFORMANCE :   Only the most remarkable performance properties are
                summarized both in the positive and negative sense.

REMARKS :   When necessary, we include some comments about
                implementation, usage, further developments, and
                explain exceptional results.

Further information about optimization software can be found in
Waren, Lasdon [WL 1979] and Nazareth [NA 1978]. A supplementary
bibliography is published by Einarsson [EI 1979].

NAME: OPRQP/XROP

AUTHOR: M.C. Bartholomew-Biggs

SOURCE: Numerical Optimisation Centre
The Hatfield Polytechnic
19 St. Albans Road
Hatfield, Herts, Great Britain

ALGORITHM: The program is classified as a quadratic approximation
method as described in more details in Section 6.1 of
Chapter II. The matrix H is obtained by the BFGS quasi-
Newton formula applied to the Lagrange function. The
penalty parameters are calculated by

$$r := 2\gamma \sqrt{g(x)^T g(x) \; v^T v_\bullet} \; , \qquad (42)$$

where $g(x)$ denotes the vector of the active constraints
and v the current approximation of the Lagrange multipliers.

VERSIONS: XROP is a revised version of OPRQP in the sense that
the constant γ in (42) is modified and that Powell's
[PO 1978a] version of the BFGS-formula is implemented.

REFERENCE: Bartholomew-Biggs [BI 1972, BI 1976, BI 1978, BI 1979]

PERFORMANCE: Both versions belong to the most efficient programs of
this study and, in particular, use the lowest calculation times. But XROP is not as reliable as OPRQP
and has lower global convergence scores. OPRQP shows
difficulties when solving ill-conditioned problems and
is much more sensitive to slight variations of the
problem than XROP which is almost as stable as penalty
methods.

REMARKS: In spite of their efficiency, the codes are programmed
using 500 statements only. As pointed out by the
author, a more reliable version of XROP is being in-
vestigated at present.

NAME: VF02AD

AUTHOR: M.J.D. Powell

SOURCE: Computer Science and Systems Division
                A.E.R.E.
                Harwell, Oxfordshire, Great Britain

ALGORITHM: The program realizes the quadratic approximation
                method of Wilson, Han, and Powell as described in
                Section 6.2 of Chapter II. The constraints in the
                quadratic subproblem (35) are modified to avoid in-
                feasibility. The steplength procedure uses quadratic
                interpolation of a non-differentiable penalty function.

REFERENCE: Powell [PO 1978a, PO 1978b]

PERFORMANCE: The outstanding advantage of VF02AD consists of its very
                low number of function and gradient evaluations. In con-
                trast to all other codes, this property does not change
                significantly when the dimension or number and type of
                the constraints of the underlying problem are raised;
                however, the calculation time seems to be correlated
                to the number of variables. In addition, VF02AD shows an
                excellent reliability and especially the lowest percentage
                of non-successful test runs. The program is not able to
                take advantage of redundant constraints. Ill-conditioned
                problems could deteriorate the efficiency.

REMARKS: The implementation of VF02AD requires the existence of a
                subroutine for the evaluation of inner products. If IBM
                360 Assembler is not available, the user has to write an
                equivalent subroutine in his own machine language. In
                contrast to all other optimization codes under considera-
                tion, the problem functions and their derivatives are
                executed together within the main program using a flag.
                No definition of any special subroutine is required.
                Although VF02AD shows an excellent efficiency, the pro-
                gram should be considered as a preliminary version because
                of some inefficiencies of the quadratic programming sub-
                routine. As pointed out by the author, some improvements
                are currently being investigated.

NAME: GRGA

AUTHOR: J. Abadie

SOURCE: J. Abadie
               University of Paris VI
               Institut de Programmation
               4, place Jussieu
               Paris, France

ALGORITHM: GRGA is Abadie's generalized reduced gradient algo-
               rithm as described in Section 7 of Chapter II. The
               search directions are obtained by the method of
               Fletcher and Reeves.

REFERENCE: Abadie [AB 1975, AB 1978], Abadie and Carpentier
               [AC 1969], Abadie and Guigou [AG 1970]

PERFORMANCE: The code is not as efficient as quadratic approxima-
               tion methods with respect to the number of function
               evaluations, but GRGA possesses an excellent relia-
               bility and global convergence. In particular, the
               feasibility of the current iterates is retained to
               a high accuracy. Degeneracy can lead to an increase
               in calculation time and function or gradient evalu-
               ations, but the program was able to solve ill-conditi-
               oned problems without any difficulties. The efficiency
               of GRGA is sensitive with respect to the position of
               the starting point.

REMARKS: GRGA can be modified to solve large problems, cf.
               Peschon and Peterson [PP 1975]. A more efficient
               version of GRGA with search directions computed by
               the BFGS-formula is developed by Abadie and Haggag
               [AH 1977, AH 1979].

NAME: OPT

AUTHOR: K.M. Ragsdell, G.A. Gabriele

SOURCE: K.M. Ragsdell
               School of Mechanical Engineering
               Purdue University
               West Lafayette, Indiana 47907, USA

ALGORITHM: The program is an implementation of a generalized
               reduced gradient method, see Section 7 of Chapter II.
               The search directions are obtained by the Fletcher-
               Reeves method and the line-search uses a quadratic
               interpolation scheme.

REFERENCE: Gabriele, Ragsdell [GR 1976], Gabriele [GA 1975]

PERFORMANCE: The performance of OPT is similar to that of GRGA
               with only one exception: GRGA is more reliable than
               OPT. Both programs compute feasible iterates in
               almost all test runs and possess the best global
               convergence scores. Degeneracy deteriorates the
               efficiency, but there are no difficulties when solving
               ill-conditioned problems. The efficiency depends sig-
               nificantly on the position of the starting point.

REMARKS: An explanation for the lower reliability and the
               higher calculation times of OPT compared with GRGA
               could be that the program does not allow the provi-
               sion of analytical derivatives for the objective
               function.

NAME: GRG2(1/2)

AUTHOR: L.S. Lasdon, A.D. Waren, M.W. Ratner

SOURCE: L.S. Lasdon
               Department of General Business
               College of Business Administration
               The University of Texas
               Austin, Texas 78712, USA

ALGORITHM: The generalized reduced gradient concept of GRG2 is
               described in detail in Section 7 of Chapter II. The
               search directions are determined by the BFGS update
               mechanism and the line-search uses a quadratic poly-
               nomial fit.

VERSIONS: GRG2(1) requires analytical first derivatives and
               GRG2(2) computes them numerically.

REFERENCE: Lasdon, Waren, Ratner [LR 1978], Lasdon, Waren
               [LW 1978], Lasdon, Waren, Jain, Ratner [LJ 1978]

PERFORMANCE: The program realizes an efficient method to solve
               nonlinear optimization problems and both versions
               belong to the most reliable codes of this study.
               Numerical differentiation requires about 50 per cent
               additional calculation time. In contrast to the other
               generalized reduced gradient algorithms, GRG2 needs
               more calculation time and a higher number of function
               or gradient evaluations when solving ill-conditioned
               problems.

REMARKS: The FORTRAN dialect of GRG2 is not as portable as
               those of other programs. Badly scaled restrictions
               could lead to irregular results. The program is well
               documented, and an extension of GRG2 to solve large,
               sparse problems is investigated, cf. Lasdon and
               Waren [LW 1978].

NAME: VF01A

AUTHOR: R. Fletcher

SOURCE: Computer Science and Systems Division
               A.E.R.E.
               Harwell, Oxfordshire, Great Britain

ALGORITHM: The program realizes Fletcher's multiplier method as
described in Section 5 of Chapter II. The unconstrained
minimizations are performed by VA09A, a quasi-Newton
method of the Harwell Subroutine Library which avoids
line-searches except on rare occasions. The update
scheme of the quasi-Newton matrices differs from
those of Section 3.2, Chapter II, in the sense that
the conjugacy property is replaced by one of monotonic
convergence of the eigenvalues, cf. Fletcher [FL 1970].

REFERENCE: Fletcher [FL 1970, FL 1975]

PERFORMANCE: VF01A is one of the most efficient multiplier methods
and even faster than most generalized reduced gradient
codes tested so far, although the program requires a
relatively high number of gradient evaluations. The
program possesses poor global convergence or, in
other words, tries to approximate the local solution
closest to the starting point. As observed for all
multiplier methods, the efficiency is improved when
all Lagrange multipliers vanish. On the other side,
ill-conditioning leads to a drastical decrease of
efficiency, and VF01A is able to solve indefinite
problems even more efficiently than definite ones.
The program is very sensitive to slight variations
of the problem.

REMARKS: The failures of VF01A are due to overflow.

NAME: LPNLP

AUTHOR: D.A. Pierre, M.J. Lowe

SOURCE: D.A. Pierre
               Electronics Research Laboratory
               Montana State University
               Bozeman, Montana 59717, USA

ALGORITHM: A special augmented Lagrange function similar to (29)
               is constructed for constraints of the form
               $g_j(x) = a_j$, $j=1,\ldots,m_e$, and $g_j(x) \geq b_j$, $j=m_e+1,\ldots,m$.
               The search directions are computed by the DFP-formula
               and a quadratic polynomial is used to approximate the
               unidirectional minimum.

REFERENCE: Pierre, Lowe [PL 1975]

PERFORMANCE: Remarkable performance properties are the attempts to
               approximate local solutions close to the starting
               point and an excellent behaviour when solving degene-
               rate problems. LPNLP is sensitive to slight variations
               of the problem, but not at all to the position of the
               starting point as observed for other multiplier
               methods like VF01A.

REMARKS: The underlying mathematical method and the program
               are well documented in the book of Pierre and Lowe
               [PL 1975]. In particular, the book contains the
               complete FORTRAN listings.

NAME: SALQDR/SALQDF/SALMNF

AUTHOR: P.E. Gill, W. Murray, S.M. Picken, E.M.R. Long

SOURCE: Division of Numerical Analysis and Computations
                 National Physical Laboratory
                 Teddington, MIDDX TW11 ODW, Great Britain

ALGORITHM: The inequality constraints are transformed to equality
                 constraints by introducing slack variables. An augmen-
                 ted Lagrange function is formulated and minimized sub-
                 ject to bounds on the original and slack variables.
                 The line-search is performed by a safeguarded step-
                 length algorithm.

VERSIONS: In the first two versions, the search directions are
                 obtained by the BFGS-formula in Cholesky decomposition.
                 SALQDR requires analytical first derivatives and SALQDF
                 computes them numerically. The third version, SALMNF,
                 uses first analytical derivatives to approximate the
                 Hessian of the Lagrangian and to perform a modified
                 Newton method.

REFERENCE: Gill, Murray [GM 1972, GM 1974, GM 1976], Murray
                 [MU 1969a, MU 1969b, MU 1976], Gill [GI 1978]

PERFORMANCE: SALQDR and SALQDF interrupt some test runs because of
                 too many calls of the unconstrained minimization
                 subroutine, leading to a lower reliability. On the
                 other hand, both programs show good global conver-
                 gence. SALQDR has difficulties in solving nearly dege-
                 nerate problems. In both cases, the use of numeri-
                 cal differentiation implies higher calculation times.
                 The main advantage of SALQDR, SALQDF, and SALMNF is
                 that they have the best ease of use scores.

REMARKS: The programs are part of the extensive and well docu-
                 mented library of the National Physical Laboratory.
                 Some failures occurred since the program could
                 not solve optimization problems with n active con-
                 straints.

NAME: CONMIN

AUTHOR: P.C. Haarhoff, J.D. Buys, H. von Molendorff

SOURCE: Atomic Energy Board
Pretoria, South Africa
(or Kuester, Mize [KM 1973])

ALGORITHM: CONMIN solves equality constrained optimization
problems by minimizing the augmented Lagrangian (26),
cf. Section 5 of Chapter II. The Lagrange multipliers
are updated by solving the equations

$$D_x g(x) D_x g(x)^T u = D_x g(x) \nabla_x f(x)$$

for each new iterate where $g(x) := (g_1(x), \ldots, g_m(x))^T$.
The unconstrained minimizations are performed by the
DFP-method with quadratic interpolation line-search.

REFERENCE: Haarhoff, Buys, von Molendorff [HB 1969], Haarhoff,
Buys [HB 1970], Kuester, Mize [KM 1973]

PERFORMANCE: The program is not reliable and has low global convergence scores. Degeneracy does not influence the
efficiency significantly, but ill-conditioned problems increase execution time and number of function
evaluations. CONMIN is sensitive to slight variations
of the problem.

REMARKS: The low reliability could be explained by the trans-
formation of the inequality to equality constraints
as described in Kuester and Mize [KM 1973]. All
problems are scaled automatically. Some failures
were due to excessive calculation times and zero
divisions. CONMIN possesses the lowest ease of use
scores. The card deck was kindly provided by Vöest-
Alpine AG (Linz, Austria).

NAME:	BIAS(1/2)

AUTHOR: K.M. Ragsdell, R.R. Root

SOURCE: K.M. Ragsdell
School of Mechanical Engineering
Purdue University
West Lafayette, Indiana 47907, USA

ALGORITHM: BIAS minimizes Rockafellar's augmented Lagrangian (29) by the DFP quasi-Newton method. The line-search procedure performs a quadratic interpolation as given by Coggins [CG 1964]. The problem is scaled initially and the variables are held between their upper and lower bounds.

VERSIONS: For BIAS(1) analytical first derivatives are required and BIAS(2) uses the embedded numerical differentiation subroutine.

REFERENCE: Root, Ragsdell [RR 1978], Root [RT 1977], Schuldt, Gabriele, Root, Sandgren, Ragsdell [SG 1977]

PERFORMANCE: The program BIAS shows average efficiency for a multiplier method. The code requires a relatively low number of gradient evaluations to solve a problem, but a high number of function calls. BIAS is quite reliable and takes no advantage of degenerate or nearly degenerate problems unless all multipliers vanish. The program solves indefinite problems even more efficiently. Numerical differentiation leads to much higher calculation times and to decreased reliability and global convergence scores. In addition, BIAS(2) shows more difficulties when solving degenerate or nearly degenerate problems, but is not as sensitive as the first version to the position of the starting point.

NAME: FUNMIN

AUTHOR: D. Kraft

SOURCE: D. Kraft
               Institut für Dynamik der Flugsysteme
               DFVLR
               8031 Oberpfaffenhofen, Germany, F.R.

ALGORITHM: The unconstrained minimizations of this multiplier
               method are performed by the BFGS quasi-Newton algo-
               rithm where the update matrices are stored in fac-
               torized form (LDL decomposition). The steplength
               algorithm is based on quadratic interpolation.

REFERENCE: Kraft [KR 1977]

PERFORMANCE: When considering the efficiency of FUNMIN, one should
               relate the corresponding scores to the achieved high
               final accuracy. The program is not very reliable and
               takes no advantage of degenerate problems unless all
               multipliers are zero. The program is sensitive to
               slight variations of the problem and in particular
               to the position of the starting point.

REMARKS: The program was originally designed to solve para-
               meterized optimal control problems, cf. Kraft
               [KR 1977], and this reference contains the complete
               FORTRAN listings.

NAME: GAPFPR

AUTHOR: D.M. Himmelblau

SOURCE: D.M. Himmelblau
Department of Chemical Engineering
The University of Texas
Austin, Texas 78712, USA

ALGORITHM: Inequality constraints are transformed to equality
constraints using slack variables. The augmented
Lagrange function (26) is minimized subject to upper
and lower bounds on the original and slack variables
and to linear constraints, if present. The search
directions are calculated by Fletcher's quasi-Newton
method [FL 1970], and Goldfarb's linear projection
technique [GF 1969] is used to satisfy all linear
constraints. During each iteration, the multipliers
are updated by one step of Broyden's rank-1-method
[BY 1967] to maximize the augmented Lagrangian with
respect to the multipliers.

REFERENCE: Newell, Himmelblau [NH 1975]

PERFORMANCE: The efficiency of GAPFPR is similar to Fletcher's
VF01A with two exceptions: VF01A is faster and can
save some gradient evaluations of the restrictions.
The reliability of GAPFPR is reduced by a relatively
high number of failures. Neither degenerate nor ill-
conditioned nor indefinite problems could change the
efficiency scores significantly. Surprisingly, the
program requires lower calculation times and less
function or gradient evaluations when starting
far away from the solution.

REMARKS: A high number of failures occurred due to excessive
calculation times and zero divisions.

NAME: GAPFQL

AUTHOR: J.S. Newell

SOURCE: D.M. Himmelblau
Department of Chemical Engineering
The University of Texas
Austin, Texas 78712, USA

ALGORITHM: The algorithm is similar to GAPFPR with the exception that the inequality constraints are incorporated in the augmented Lagrange function as proposed by Rockafellar, cf. Section 5 of Chapter II.

REFERENCE: Newell, Himmelblau [NH 1975]

PERFORMANCE: The overall efficiency and reliability of GAPFQL are comparable with the results obtained by GAPFPR. But in contrast to the latter code, GAPFQL computes an extremely low percentage of global solutions, or, in other words, tries to approximate a local solution closest to the starting point. The performance for solving degenerate or ill-conditioned problems is very similar to other multiplier methods in the sense that the program takes advantage of redundant constraints and shows difficulties when solving ill-conditioned problems. The solution of indefinite problems leads to increased efficiency scores.

REMARKS: Many failures occurred due to overflow and excessive calculation times. The implementation of GAPFQL requires only 450 FORTRAN statements.

NAME: ACDPAC

AUTHOR: M.J. Best, A.T. Bowler

SOURCE: M.J. Best
                Department of Combinatorics and Optimization
                University of Waterloo
                Waterloo, Ontario N2L 3G1, Canada

ALGORITHM: ACDPAC minimizes Rockafellar's augmented Lagrange
                function subject to bounds on the variables and
                linear constraints if present, by the accelerated
                conjugate direction method of Best and Ritter
                [BR 1976]. Conjugate directions are obtained by
                choosing search directions orthogonal to differences
                of gradients at previous iterates. The line-search
                procedure uses the Armijo test (9) as described in
                Section 1.1 of Chapter II, but it can be shown
                [BR 1976] that the steplength converges to one.
                In addition, the Best-Ritter algorithm possesses
                a superlinear rate of convergence, cf. Best and
                Ritter [BR 1976].

REFERENCE: Best, Bowler [BB 1978], Best, Ritter [BR 1975, BR 1976]

PERFORMANCE: ACDPAC is a very reliable program and in particular
                the most reliable multiplier method. The most out-
                standing result, however, is the worst efficiency of
                all programs when solving degenerate or nearly dege-
                nerate problems. The remaining performance properties
                are similar to other multiplier methods.

NAME: FMIN(1/2/3)

AUTHOR: F. Lootsma, D. Kraft

SOURCE: D. Kraft
              Institut für Dynamik der Flugsysteme
              DFVLR
              8031 Oberpfaffenhofen, Germany, F.R.

ALGORITHM: FMIN is an implementation of a penalty method and is
              based on minimizing an interior-exterior penalty function.
              The unconstrained minimizations are either performed by
              the BFGS or a modified Newton algorithm. The line-search
              procedure uses successive quadratic interpolations.

VERSIONS: FMIN(1) requires analytical derivatives for the BFGS
              method, FMIN(2) computes them numerically, and FMIN(3)
              needs analytical first derivatives to realize a modified
              Newton method by numerical approximations of the Hessians.

REFERENCE: Lootsma [LO 1974, LO 1978]

PERFORMANCE: All tested penalty methods are relatively inefficient,
              since the growth of the penalty parameter generates in-
              creasingly ill-conditioned unconstrained programs.
              Nevertheless, FMIN(1) needs as many gradient calls of the
              objective function as VF01A. The numerical differentiation
              in FMIN(2) leads to higher calculation times and to a
              lower reliability. The third version realizing Newton's
              method has better global convergence and reliability and
              uses a lower number of unconstrained minimizations, but,
              on the other hand, calculation time and number of gradient
              evaluations are raised drastically. FMIN(1) does not change
              its efficiency significantly when solving degenerate, ill-
              conditioned, or indefinite test problems indicating that
              the logarithmic barrier term of the penalty function de-
              stroys the curvature of f inside the feasible region. It
              turns out that FMIN(1) is not sensitive to the position of
              the starting point. The program uses a higher number of
              function calls when starting far away from the solution,
              but an even slightly reduced number of iteration steps.

REMARKS: The method was first developed and implemented in ALGOL
              60 by Lootsma [LO 1978]. Subsequently, the program was
              translated into FORTRAN by D. Kraft.

NAME: NLP

AUTHOR: D. Rufer

SOURCE: D. Rufer
Fachgruppe für Automatik
ETH - Zentrum
8092 Zürich, Switzerland

ALGORITHM: The algorithm minimizes the exterior penalty function
(23) by the quasi-Newton method of Fletcher [FL 1970]
combined with a golden section line-search.

REFERENCE: Rufer [RU 1978]

PERFORMANCE: As observed for all penalty methods, NLP is not very
efficient, but quite reliable despite a relatively
high number of failures due to excessive calculation
times. The program could solve nearly degenerate
problems even more efficiently and is not sensitive
both to slight variations of the problem and to the
position of the starting point.

REMARKS: The program is well documented and part of a major
subroutine package which includes algorithms for
solving nonlinear identification problems, cf. Rufer
[RU 1979].

NAME: SUMT

AUTHOR: W.C. Mylander, R.L. Holmes, G.P. McCormick

SOURCE: Research Analysis Corporation
             McLean, Virginia, USA
             (or Kuester, Mize [KM 1973])

ALGORITHM: First, the algorithm determines a feasible solution
             and minimizes an interior penalty function (24) for
             a sequence of increasing penalty parameters. The
             unconstrained minimizations are performed by a modi-
             fied DFP-procedure with golden section line-search.

REFERENCE: Mylander, Holmes, McCormick [MH 1971], Fiacco,
             McCormick [FM 1968], Kuester, Mize [KM 1973]

PERFORMANCE: The program is one of the earliest codes to solve
             nonlinear optimization problems in an efficient way
             and is included in our test series to allow compari-
             sons with older studies and to show the progress
             which has been made in the last years to improve the
             efficiency of optimization programs. The major draw-
             backs of SUMT are high calculation times and numbers
             of function evaluations resulting from an inefficient
             line-search. In addition, the program is unreliable
             and has low global convergence scores. SUMT is not
             at all sensitive to slight variations of the problem
             and, in particular, one cannot observe any signifi-
             cant change of the efficiency items when solving
             degenerate, ill-conditioned, or indefinite problems.
             In addition, the program is not sensitive to the
             position of the starting point.

REMARKS: Many test runs are interrupted by an internal
             time limit since the convergence conditions do not
             work satisfactorily, and some failures are caused
             by overflow. The card deck was kindly provided by
             Vöest-Alpine AG (Linz, Austria).

NAME: DFP

AUTHOR: J.P. Indusi

SOURCE: Computing Center
              S.U.N.Y.
              Stony Brook, New York, USA

ALGORITHM: After transforming the inequality constraints to
              equality constraints, an exterior penalty function
              of the type (23) is formulated. For a sequence of
              increasing penalty parameters this function is mini-
              mized by a modified DFP-algorithm with a combined
              golden section and cubic interpolation steplength
              procedure. In the neighbourhood of a solution, a
              Newton-type acceleration phase is performed.

REFERENCE: Indusi [IN 1972]

PERFORMANCE: DFP belongs to the most efficient penalty methods
              and has a good global convergence behavior; how-
              ever, it is one of the most unreliable nonlinear
              programming codes. Ill-conditioning in the problem
              leads to a drastical deterioration of efficiency.
              The program is very sensitive to slight variations
              of the problem and the position of the starting
              point and, in addition, has low ease of use scores.

REMARKS: The program could not solve 34 test problems of
              80 in the first test series. Besides some test
              runs interrupted by excessive calculation times or
              overflow, most of these failures occurred since
              the program could not proceed to the acceleration
              phase because of rank-deficiency. This leads to
              infinite loops although the iterates are close to the
              solution.

NAME: FCDPAK

AUTHOR: M.J. Best

SOURCE: M.J. Best
        Department of Combinatorics and Optimization
        University of Waterloo
        Waterloo, Ontario N2L 3G1, Canada

ALGORITHM: The version of FCDPAK used solves inequality constrained nonlinear programs by Robinson's method as
described in Section 8 of Chapter II. The linearly
constrained subproblems are solved by the accelerated
conjugate direction method of Best and Ritter
[BR 1975]. This method is similar to that used in
ACDPAC, cf. Best and Ritter [BR 1976].

REFERENCE: Best [BS 1975a, BS 1975b], Best, Ritter [BR 1975],
Robinson [RB 1972]

PERFORMANCE: The program belongs to the most efficient of
this comparative study. In particular, only OPRQP
and XROP are faster and only VF02AD requires a lower
number of gradient evaluations of the restrictions.
The reliability is reduced by some test runs for
which the stopping criteria are met during the first
iteration. FCDPAK possesses a good global convergence
and profits from degenerate or nearly degenerate
problems. Ill-conditioned problems are solved as
efficiently as by the generalized reduced gradient
methods OPT and GRGA, but a drawback of the program
is the extremly high sensitivity to the position of
the starting point.

REMARKS: Equality constraints are transformed to inequality
constraints by $g_j(x) \le \epsilon$, $g_j(x) \ge -\epsilon$ for $j=1,\ldots,m_e$
and $\epsilon := {}_{10}-4$. The zero tolerances of the program are
not adaptable to another machine precision. In spite
of some drawbacks of this implementation, Robinson's
method should get more attention for future research
in particular since this method seems to be able to
solve large, sparse problems.

THE CONSTRUCTION OF TEST PROBLEMS

Any development or comparison of nonlinear programming software
has to be based on extensive numerical tests. These depend on test
problems of the form (1), implemented in an appropriate way, about
whose mathematical structure as much as possible is known. Most
test problems used in the past to test and compare optimization
programs consist of so called 'real life' problems which are believed to reflect typical structures of practical nonlinear programming problems, for example the Colville [CO 1968] problems, cf.
Himmelblau [HI 1972] or Hock, Schittkowski [HS 1980]. But this class
of test examples has some severe disadvantages especially since
their precise solution is not a priori known so that the efficiency
of a code cannot be related to the achieved accuracy. In this chapter, a completely different approach is presented: The construction
of randomly generated test problems with predetermined solutions.

1. Fundamentals of the test problem generator

A test problem generator is presented satisfying the following
conditions:

a) It is possible to produce several classes of test problems such as
   small and dense problems, big and sparse problems, problems with
   equality or inequality constraints only, etc..

b) Each class of test problems is completely described by very few
   parameters, e.g., dimension, number of constraints, upper
   and lower bounds.

c) A repeated execution of the generator yields arbitrarily many
   different problems of the class randomly.

d) An optimal solution, i.e. a point satisfying the Kuhn-Tucker and
a second order condition, is known a priori. The corresponding
precise objective function value is zero.

e) It is possible to construct test problems with special properties,
e.g. convex, linearly constrained, ill-conditioned, degenerate,
or indefinite problems.

f) Each problem can be provided with different randomly generated
starting points.

Indeed, a test problem generator on these lines will provide
wide range of different problems for general purpose tests as well
as problems with special features. To realize the above requirements,
let s_j, $j=0,1,\ldots,m$, be any series of twice differentiable functions
defined on a subset of $\mathbf{R}^n$ containing the interval $\{x \in \mathbf{R}^n : x_1 \le x \le x_u\}$.
Furthermore we need a randomly chosen $x^* \in \mathbf{R}^n$ with

$$x_1 < x^* < x_u$$

which will define a (at least) local minimizer of the optimization
problem.

First we must establish that x^* is feasible, with exactly m_a
active constraints, where m_a is a predetermined integer. In addition,
it should be possible for constructing special types of test problem
to predetermine the gradients $\nabla_x g_j(x^*)$, $j=1,\ldots,m_e+m_a$. We therefore
define the restrictions by

$$g_j(x) := s_j(x) - s_j(x^*) + d_j^T(x^* - x) , \quad j=1,\ldots,m_e+m_a$$
$$g_j(x) := s_j(x) - s_j(x^*) + \mu_j , \quad j=m_e+m_a+1,\ldots,m, \tag{43}$$

where $d_j \in \mathbf{R}^n$, $j=1,\ldots,m_e+m_a$, and the real numbers μ_j, $j=m_e+m_a+1,\ldots,m$,
are randomly chosen within the interval $(0,m)$.

The objective function f is defined by

$$f(x) := s_0(x) + \frac{1}{2} x^T H x + q^T x + \alpha \tag{44}$$

with an (n,n) matrix H, $q \in \mathbf{R}^n$, and $\alpha \in \mathbf{R}$. The quadratic term of f
has to be determined so that x^* satisfies the Kuhn-Tucker condition
(3), the second order condition (4), and the condition $f(x^*) = 0$.
Therefore we determine optimal Lagrange multipliers u_j^*, $j=1,\ldots,m$,

with

$$u_j^* \geq 0 , \quad j=m_e+1,\ldots,m_e+m_a,$$
$$u_j^* = 0 , \quad j=m_e+m_a+1,\ldots,m,$$

(45)

furthermore an (n,n) matrix P with

$$y^T P y > 0$$

(46)

for all nonzero vectors y with $y^T \nabla_x g_j(x^*) = 0$, $j=1,\ldots,m_e$, and $y^T \nabla_x g_j(x^*) = 0$ for all j with $u_j^* > 0$, $j=m_e+1,\ldots,m$. It is easy to see that the definition of the matrix

$$H := -\nabla_x^2 s_0(x^*) + \sum_{j=1}^{m} u_j^* \nabla_x^2 g_j(x^*) + P$$

(47)

leads to $\nabla_x^2 L(x^*,u^*) = P$ implying that the second order condition (4) is always satisfied. The Kuhn-Tucker condition $\nabla_x L(x^*,u^*) = 0$

$$q := -\nabla_x s_0(x^*) - Hx^* + \sum_{j=1}^{m} u_j^* \nabla_x g_j(x^*) .$$

(48)

Finally, the constant term α is given by

$$\alpha := -s_0(x^*) - \frac{1}{2} x^{*T} H x^* - q^T x^*$$

(49)

and guarantees that $f(x^*) = 0$. This completes the construction of a test problem provided that one knows how to choose the following data:

a) The series of twice continuously differentiable functions $s_0,\ldots,s_m$.

b) The linear terms of the restrictions, i.e. the vectors $d_1,\ldots,d_{m_e+m_a}$.

c) The optimal Lagrange multipliers, i.e. any $u^* = (u_1^*,\ldots,u_m^*)^T$ satisfying (45).

d) The Hessian of the Lagrangian with respect to x^* and u^*, i.e. a matrix P satisfying (46).

These data are specialized in the following sections to allow the construction of optimization problems in accordance with the indi-vidual purposes of the test designer. Numerical experience showed that scaling problems might be created since the computed numbers de-termining H, q, and α are sometimes several orders of magnitude larger than the input values. To avoid numerical difficulties, H, q, and α should be evaluated using double precision arithmetic, whereas a driving program executing an optimization code could be written in single precision. This ensures that the test problem generator has more exact arithmetic than the codes which are to be tested. Neverthe-

less, one should check numerically for each problem whether x* defines
a local minimizer. In Appendix B, a sensitivity analysis is presented
for all test problems which are constructed to obtain the numerical
results of Chapter V, showing whether one could accept the predetermined x* as a solution of the computed test problem.

2. General test problems

We consider now the construction of test problems for general
purpose tests, i.e. for tests determining the overall efficiency,
reliability, and global convergence of an optimization program. In
this case, one could define the functions $s_o, s_1, \ldots, s_m$ by signomials,
i.e. generalized polynomial functions of the kind

$$s(x) = \sum_{j=1}^{k} c_j \prod_{i=1}^{n} x_i^{a_{ij}} , \quad x > 0 , \tag{50}$$

where the coefficients c_j and the exponents a_{ij} are real numbers.
Functions of this kind are considered because of their simple structure and the observation that many 'real life' problems are defined
by signomials, for example chemical equilibrium problems, cf. Duffin,
Peterson, Zener [DP 1967]. Since each signomial is completely described
by the data c_j and a_{ij}, j=1,...,k, i=1,...,n, it is possible to produce these data randomly using predetermined bounds. In accordance
with 'real life' geometric programming problems which are defined by
functions of the kind (50), it should be possible to have integer exponents for the signomials, cf. Dembo [DE 1976a]. Furthermore one
should implement the possibility to vary the density of the coeffici-
ent matrix $(a_{ij})_{i,j=1,n}$. Since s(x) is only defined for positive variables, one could expand s by any convex function outside of the posi-
tive cone, i.g., by $s(x):= \sum_{i=1}^{n} (x_i - e_i)^2 + \eta$ with $e_i > 0$ and $\eta \gg 0$.

Since it is not required to predetermine the gradients of the
active constraints, we let $d_j = 0$, j=1,...,$m_e + m_a$. The optimal Lagrange
multipliers u_j^* are given by the instructions

$$u_j^* \in (b_1, b_2) \text{ randomly chosen, } j=1,\ldots,m_e,$$

$$u_j^* \in (0, b_3) \text{ randomly chosen, } j=m_e+1,\ldots,m_e+m_a, \qquad (51)$$

$$u_j^* = 0, \quad j=m_e+m_a+1,\ldots,m.$$

To satisfy the sufficient second order optimality condition, consider an upper triangular matrix U with elements randomly chosen in the interval (b_4, b_5) and compute the positive definite matrix

$$P := U^T U . \qquad (52)$$

These definitions satisfy the requirements of the last section for the construction of a test problem. The bounds $b_1, \ldots, b_5$ for determining the Lagrange multipliers and the elements of U are predetermined by the user. The reader should be aware that in general the signomials are not convex functions implying that the given solution x^* is only a local one. In other words, it is possible that an optimization code approximates a solution with an objective function value less than zero. This is not considered to be a disadvantage, since these test runs can be used to determine the global convergence of an optimization program, see Chapter V. Appendix A contains the data for constructing 80 test problems as described in this section and, in addition, a detailed example.

3. Linearly constrained test problems

Test problems with linear equaliy and inequality constraints are easily obtained by defining

$$s_j(x) := a_j^T x , \quad j=1,\ldots,m,$$

with randomly chosen vectors $a_j \in \mathbb{R}^n$, $j=1,\ldots,m$. If one is not required to predetermine the gradients of the active constraints, let $d_j = 0$ for $j=1,\ldots,m_e+m_a$. For the construction of the objective function f, one could use any signomial s_0 of the kind (50), furthermore the instructions (51) and (52) for determining the Lagrange multipliers u* and the Hessian $\nabla_x^2 L(x^*, u^*)$.

In addition, it is possible to generate convex linearly constrained test problems. In this case, one could replace s_o by a convex exponential sum of the form

$$s(x) = \sum_{j=1}^{k} c_j \exp(\sum_{i=1}^{n} a_{ij} x_i) \tag{53}$$

with randomly generated $c_j > 0$ and $a_{ij} \in \mathbf{R}$. The optimal Lagrange multipliers are given by (51). To guarantee the convexity of f on $\mathbf{R}^n$, one could set H = 0 or, equivalently, $P = \nabla_x^2 s_o(x^*)$. This matrix is at least positive semi-definite and positive definite, if a positive definite matrix P' is added to P, i.e. if

$$P := \nabla_x^2 s_o(x^*) + P' \; ,$$

so that all assumptions for the construction of a test problem are satisfied. In the convex case, the local minimizer is a global one.

4. Degenerate test problems

An optimization problem of the kind (1) is called degenerate, if at least one of the Lagrange multipliers u_j^*, $j=1,\ldots,m_e+m_a$, vanishes, i.e. degeneracy occurs if at least one of the active constraints is redundant at the optimal solution x*. If in the worst case all Lagrange multipliers are zero, the constrained local minimizer x* is identical with an unconstrained local minimizer of f. We denote a test problem nearly degenerate, if the Lagrange multipliers differ widely in their order of magnitude. Both situations arise in practical applications and by numerical experiment we try to get an answer to the question: How does an optimization code behave under different degrees of degeneracy. In particular, we are interested in the following questions:

a) Are there any numerical difficulties when solving degenerate problems?

b) Does an optimization program take any advantage of redundant constraints or nearly degenerate problems?

c) Does degeneracy influence the final accuracy of an optimization program?

Proceeding from a set of signomials s_j, $j=0,1,\ldots,m$, and $d_j = 0$, $j=1,\ldots,m_e+m_a$, the matrix P could be determined by (52) guaranteeing the second order condition. Test problems with varying degree of degeneracy are obtained for example by the following four conditions:

$$
\begin{array}{lll}
\text{a)} & u_j^* = 1 , & j=1,\ldots,m_e+m_a. \\[4pt]
\text{b)} & u_j^* = 10^{(-2\cdot(j-1))} , & j=1,\ldots,m_e+m_a. \\[4pt]
\text{c)} & u_j^* = 1 , & j=1,\ldots,[\tfrac{1}{2}(m_e+m_a)], \\[4pt]
& u_j^* = 0 , & j=[\tfrac{1}{2}(m_e+m_a)]+1,\ldots,m_e+m_a. \\[4pt]
\text{d)} & u_j^* = 0 , & j=1,\ldots,m_e+m_a.
\end{array} \tag{54}
$$

To allow for comparison of the results, the test problems should be distinguished only by these Lagrange multipliers. All other data such as dimension, number of constraints, signomials s_j, $j=0,1,\ldots,m$, etc. should be identical, leading to a series of test problems with an increasing degree of degeneracy. Appendix A contains the data for the construction of 8 test problems in each of the four cases (54).

5. Ill-conditioned test problems

It is outlined in Theorem 2 of Chapter 1 that at least for convex problems any saddle point of the Lagrangian defines a minimizer of (1). In this case, the solution x* is a minimizer of the function $L(x,u^*)$, where u* denotes the optimal Lagrange multipliers. Numerical experience in unconstrained optimization shows that the local convergence of a standard unconstrained nonlinear programming code depends heavily on the condition number of the Hessian matrix at the optimal solution, in our case on $\text{cond}(\nabla_x^2 L(x^*,u^*)) = \text{cond}(P)$. Since many programs designed for the solution of the constrained problem are based on minimizing an augmented Lagrangian, we intend to construct test problems with different condition numbers of the Hessian matrix $\nabla_x^2 L(x^*,u^*)$ and we are concerned with the question how ill-conditioning influences the final accuracy and efficiency (execution time and number of function and gradient evaluations) of an optimization code.

For generating ill-conditioned test problems, we use a set of signomials s_j, $j=0,1,\ldots,m$, furthermore $d_j = 0$ for $j=1,\ldots,m_e+m_a$, and randomly chosen multipliers u^*, see (51). The matrix P is defined by

$$P := \begin{pmatrix} H_\nu & : & 0 \\ 0 & : & I_{n-\nu} \end{pmatrix} \tag{55}$$

where $I_{n-\nu}$ denotes the $(n-\nu,n-\nu)$ unit matrix and H_ν the (ν,ν) positive definite Hilbert matrix

$$H_\nu := \left(\frac{1}{i+j-1} \right)_{i,j=1,\nu} .$$

It is obvious that the sufficient optimality criteria are satisfied. By varying ν it is possible to produce test problems in which the condition number of the Hessian of the Lagrangian increases. This condition number is approximately $\exp(3.5 \cdot \nu)$, cf. Zielke [ZI 1974], and to give an example, consider $\nu = 3,5,8$:

$$\text{cond}(H_3) \approx 3.6_{10}4,$$
$$\text{cond}(H_5) \approx 4.0_{10}7, \tag{56}$$
$$\text{cond}(H_8) \approx 1.4_{10}12.$$

To obtain the numerical results of the subsequent chapter, these matrices are used to construct 24 ill-conditioned test problems. The data are given in Appendix A.

6. Indefinite test problems

So far, we have assumed that the Hessian matrix $\nabla_x^2 L(x^*,u^*)$ is positive definite. This is a stronger assumption than required by the second order condition (4) and not always satisfied in practice. Therefore, we intend to construct indefinite test problems to check if an indefinite Hessian of the Lagrangian leads to numerical difficulties, to a different final accuracy, or to an increased efficiency.

First we again define a set of signomials s_j, $j=0,1,\ldots,m$, and the linear terms of the restrictions are given by

$$d_j := \nabla_x s_j(x^*) - e_j , \quad j=1,\ldots,m_e+m_a , \tag{57}$$

where e_j denotes the j-th axis vector. As a consequence, the gradients of the first m_e+m_a restrictions at x^* are axis vectors, i.e.

$$\nabla_x g_j(x^*) = e_j , \quad j=1,\ldots,m_e+m_a ,$$

cf. (43). The Lagrange multipliers u_j^* are randomly chosen as described by (51) with the additional assumption that $u_j^* \neq 0$, $j=1,\ldots,m_e+m_a$. The matrix P is given by

$$P := \begin{pmatrix} \sigma U_1^T U_1 & : & 0 \\ 0 & : & U_2^T U_2 \end{pmatrix} \begin{array}{l} \} \quad m_e+m_a \\ \} \quad n-m_e-m_a \end{array}$$

with upper triangular matrices U_1 and U_2, whose elements are randomly chosen between predetermined bounds. The matrix P is indefinite if and only if $\sigma \leq 0$. For a

$$y := \begin{pmatrix} z_1 \\ z_2 \end{pmatrix}, \quad z_1 \in \mathbb{R}^{m_e+m_a} , \quad z_2 \in \mathbb{R}^{n-m_e-m_a} , \quad y \neq 0 ,$$

the condition $y^T \nabla_x g_j(x^*) = 0$ for $j=1,\ldots,m_e+m_a$, is equivalent with $z_1 = 0$ leading to

$$y^T \nabla_x^2 L(x^*,u^*)y = z_2^T U_2^T U_2 z_2 > 0 .$$

This implies the validity of the second order sufficient condition for all values of σ. The numerical data for the construction of 24 test problems with $\sigma < 0$, $\sigma = 0$, or $\sigma > 0$ are presented in Appendix A.

7. Convex test problems

An optimization problem (1) is called convex if the objective
function f and the set of all feasible points are convex. The last
condition is satisfied if there are no equality constraints ($m_e = 0$)
and if all restriction functions g_j, j=1,...,m, are concave. The main
attribute of convex problems is the fact that every local minimizer
is a global one preventing difficulties with alternate solutions.
One can thus construct test problems with a starting point far away
from the solution x*. Solving the same problem with a starting point
close to the minimizer x*, gives information on the sensitivity of an
optimization program with respect to the position of the starting
point. Furthermore, it is possible to test if a code is able to take
advantage of the convex structure of an optimization problem.

First we have to look for a method to generate convex functions.
One possible way is to define exponential sums of the form (53) with
randomly chosen $c_j > 0$ and $a_{ij} \in \mathbb{R}$, j=1,...,k, i=1,...,n. It is easy
to see that these functions are derived from signomials by simple
exponential transformations $y_i = \exp(x_i)$, i=1,...,n. Consider now
m+1 convex exponential sums $t_0, t_1, ..., t_m$ and define

$$s_0(x) := t_0(x) \ ,$$
$$s_j(x) := -t_j(x) \ , \quad j=1,...,m.$$

Using the functions s_j, j=0,1,...,m, and the instructions of Section 1,
we get concave restrictions g_j, j=1,...,m. In this case, we may set
$d_j = 0$, j=1,...,m_a. The optimal Lagrange multipliers are randomly
chosen positive numbers and to achieve H = 0, define

$$P := \nabla_x^2 s_0(x^*) - \sum_{j=1}^{m} u_j^* \nabla_x^2 g_j(x^*) \ .$$

This matrix is positive semi-definite, since s_0 is convex, since $u_j > 0$,
and since the functions g_j, j=1,...,m, are concave. To force P to be
positive definite, one could add a positive definite matrix P' to P
leading to a strictly convex objective function. This completes the
construction of a convex test problem with a predetermined global
minimizer. The numerical data for defining a set of 25 problems are
presented in Appendix A.

Chapter V . . . .

PERFORMANCE EVALUATION

In this chapter, we explain nine performance criteria, describe
their evaluation, and present numerical results as a basis for comparing optimization programs from various points of view. In addition,
we show by an example, how a decision maker could obtain a final
score for each program depending on the individual significance of
the criteria.

1. Notation

For any termination point $\bar{x} \in \mathbb{R}^n$ of a test run we have to decide
if $\bar{x}$ approximates the predetermined solution x^* or not. This decision
is based on the following quantities:

Objective function value:
$$f(\bar{x})$$

Sum of constrained violations:
$$r(\bar{x}) := \sum_{j=1}^{m_e} |g_j(\bar{x})| \quad - \sum_{j=m_e+1}^{m} \min(0, g_j(\bar{x}))$$

Norm of Kuhn-Tucker vector: (58)
$$h(\bar{x}) := \|\nabla_x L(\bar{x}), u^*)\|_2$$

Number of exact digits:
$$e(\bar{x}) := \frac{1}{n} \sum_{i=1}^{n} \gamma_i$$

$$\text{with} \quad \gamma_i := \begin{cases} -\log_{10} |(\bar{x}_i - x_i^*)/x_i^*|, & \bar{x}_i \neq x_i^*, \; x_i^* \neq 0 \\ -\log_{10} |\bar{x}_i| & , \; \bar{x}_i \neq 0, \; x_i^* = 0 \\ 12 & , \text{otherwise.} \end{cases}$$

In general, there is no guarantee that x* defines a global minimizer of the corresponding optimization problem. It is therefore possible that $\overline{x}$ will sometimes approximate a global minimizer with an objective function value less than zero. This leads to three categories of test runs:

    𝔖 : Set of all successful solutions.

    𝔑 : Set of all non-successful solutions.

    𝔊 : Set of all global solutions.

Whether, in each case, $\overline{x}$ belongs to 𝔖, 𝔑, or 𝔊, depends on the special design of the test series and is explained in the following sections. Proceeding from a given separation of a battery of test runs into the sets 𝔖, 𝔑, and 𝔊, we are now able to define the performance measures accuracy, efficiency, efficiency related to accuracy, reliability, and global convergence.

a) <u>Accuracy</u>: To evaluate the achieved accuracy, we consider only successful test runs.

FV : Geometric mean of the absolute objective function values, i.e. of $|f(\overline{x})|$, $\overline{x} \in$ 𝔖.

VC : Geometric mean of the sums of constraint violations, i.e. of $r(\overline{x})$, $\overline{x} \in$ 𝔖.

KT : Geometric mean of the Euclidean norms of Kuhn-Tucker vectors, i.e. of $h(\overline{x})$, $\overline{x} \in$ 𝔖.

ED : Arithmetic mean of the numbers of exact digits, i.e. of $e(\overline{x})$, $\overline{x} \in$ 𝔖.

b) <u>Efficiency</u>: The efficiency is measured by execution time and number of function and gradient calls. In all cases, the arithmetic mean is used with respect to the successful runs.

ET : Mean value of the execution time in seconds.

NF : Mean value of the function calls of f.

NG : Mean value of the function calls of the constraints (Each restriction counted).

NDF : Mean value of the gradient calls of f.

NDG : Mean value of the gradient calls of the constraints (Each restriction counted).

c) <u>Efficiency related to accuracy</u>: To relate the efficiency to the achieved accuracy, we denote by

$$a(\overline{x}) := \frac{1}{4} \left(-\log_{10}|f(\overline{x})| - \log_{10}r(\overline{x}) - \log_{10}h(\overline{x}) + e(\overline{x}) \right) \quad (59)$$

the accuracy of a successful test run $\overline{x} \in \mathfrak{S}$. The efficiency items execution time and number of function or gradient evaluations are divided by $a(\overline{x})$ for all $\overline{x} \in \mathfrak{S}$ and subsequently the corresponding arithmetic mean values are evaluated.

ET/A : Mean value of execution time related to accuracy.

NF/A : Mean value of the objective function calls related to accuracy.

NG/A : Mean value of the constraint function calls related to accuracy (Each restriction counted).

NDF/A : Mean value of the gradient calls of f related to accuracy.

NDG/A : Mean value of the gradient calls of the constraints related to accuracy (Each restriction counted).

d) <u>Reliability</u>: Besides the percentage of non-successful test runs, the average objective function values and sums of constraint violations of the non-successful solutions are presented. It is then possible to distinguish between attempts to reach the solution x* and divergence of the program.

PNS : Percentage of non-successful solutions, i.e.
         PNS := $|\mathfrak{N}| \cdot 100/(|\mathfrak{S}| + |\mathfrak{N}| + |\mathfrak{O}|)$.

FFV : Geometric mean of the positive objective function values, i.e. of $\max(0, f(\overline{x}))$, $\overline{x} \in \mathfrak{N}$.

FVC : Geometric mean of the sums of constraint violations, i.e. of $r(\overline{x})$, $\overline{x} \in \mathfrak{N}$.

F   : Number of problems which could not be solved due to excessive calculation time (> 10 min), overflow, division by zero, etc..

e) <u>Global convergence</u>: To give more insight into global convergence behavior, the percentage of global solutions is listed along with the corresponding objective function values. One can thus distinguish between failures in approximating x* and solutions below zero.

PGS : Percentage of global solutions, i.e.
$$PGS := |\mathfrak{G}| \cdot 100 / (|\mathfrak{S}| + |\mathfrak{G}|).$$

GFV : Geometric mean of the objective function values, i.e.
of $f(\overline{x})$, $\overline{x} \in \mathfrak{G}$.

GVC : Geometric mean of the sums of constraint violations, i.e.
of $r(\overline{x})$, $\overline{x} \in \mathfrak{G}$.

2. Efficiency, reliability, and global convergence

First we will have to decide whether a termination point considered a successful, global, or non-successful solution. We define these categories in the following way:

$\overline{x} \in \mathfrak{S}$ if and only if $|f(\overline{x})| < \eta_1$, $r(\overline{x}) < \eta_2$, $h(\overline{x}) < \eta_3$, $e(\overline{x}) > \eta_4$.

$\overline{x} \in \mathfrak{G}$ if and only if $f(\overline{x}) \leq -\eta_1$, $r(\overline{x}) < \eta_2$
$\qquad\qquad$ or $f(\overline{x}) < 0$, $r(\overline{x}) < \eta_2$, $(h(\overline{x}) \geq \eta_3$ or $e(\overline{x}) \leq \eta_4)$.

$\overline{x} \in \mathfrak{N}$ otherwise.

For evaluating the results of this section, the tolerances $\eta_1, \ldots, \eta_4$ are set to

$$\eta_1 := .001 \; , \quad \eta_2 := .001 \; , \quad \eta_3 := .1 \; , \quad \eta_4 := 2 \; . \qquad (60)$$

These tolerances may influence the final scores significantly. We therefore present an analogous performance evaluation for another choice of the parameters (60) in Appendix C. Their definition is not related to the individual test problems since most of them are well scaled.

Based on the prescriptions of Section 1 and 2 of Chapter IV and the numerical data presented in Appendix A, we constructed 80 test problems with 3 different starting points, respectively. The numerical results of the 240 test runs are summarized in Tables 2 to 6. In detail, Table 2 presents the final accuracy, Table 3 the efficiency, Table 4 the efficiency related to accuracy, Table 5 the reliability, and Table 6 the global convergence of the optimization programs. The five items determining efficiency related to accuracy are scaled in such a way that the best code obtains the value 1.

We therefore denote them by ET/A/B, NF/A/B, NG/A/B, NDF/A/B, and
NDG/A/B, respectively. If NDF = 0, or NDG = 0, we set NDF = 1 and
NDG = 1. Part a of Tables 3 to 6 contains the original results de-
scribed so far and part b the corresponding consecutive rank numbers.

To give a more comprehensive impression of the performance of an
optimization program, it is possible to evaluate scores for the
criteria efficiency related to accuracy, reliability, and global
convergence by average rank numbers. Since the items determining
one performance criterium have different importance for a decision
maker, one should scale them in an appropriate way. For example,
consider the weights of Table 7 which are obtained by multicriteria
analysis as described in Appendix D in more detail. The number of
gradient calls got a higher weight than the number of function calls
since it is in general more expensive to calculate gradients than
function values. Since the importance of calculation time versus
number of function or gradient evaluations varies in practical
applications, we present two different combinations of weights for
the efficiency items. This provides for situations in which the
execution time is more or less important than the number of function
and gradient calls. The corresponding weighted average rank numbers
of the optimization programs are presented in Table 8 using the
notation for performance criteria defined in Table 7. However,
these results depend on the individual estimation of the importance
of the performance criteria and a decision maker is urged to
choose his own weights in the case of disagreement when selecting
a code.

More detailed results are listed in Appendix C where for each
optimization program, the average performance results of all classes
are reported. There an interested reader will find information
about efficiency, reliability, and global convergence for the solution
of problems with equality or inequality constraints only, with many
or few active constraints, with dense or sparse Hessians of the
signomials, etc.. In addition, will find data on the performance
of the optimization programs for solving problems with increasing
dimensions or number of constraints.

Code	FV	VC	KT	ED
OPRQP	.11E-7	.51E-8	.41E-6	7.31
XROP	.68E-7	.16E-7	.10E-4	6.10
VFO2AD	.36E-8	.40E-10	.35E-5	6.64
GRGA	.84E-5	.30E-11	.75E-3	4.37
OPT	.25E-4	.14E-8	.33E-2	3.64
GRG2(1)	.48E-6	.62E-7	.37E-4	5.77
GRG2(2)	.14E-5	.97E-8	.72E-3	4.52
VFO1A	.43E-8	.13E-8	.62E-6	7.26
LPNLP	.93E-6	.14E-5	.36E-4	5.00
SALQDR	.62E-4	.14E-5	.68E-2	3.13
SALQDF	.70E-4	.58E-6	.72E-2	3.11
SALMNF	.82E-7	.68E-7	.14E-4	5.65
CONMIN	.78E-6	.19E-7	.27E-3	4.75
BIAS(1)	.24E-5	.16E-5	.84E-4	4.79
BIAS(2)	.10E-4	.40E-5	.12E-2	3.71
FUNMIN	.65E-9	.44E-9	.88E-5	6.18
GAPFPR	.10E-4	.26E-10	.23E-2	3.76
GAPFQL	.11E-5	.73E-6	.17E-4	5.13
ACDPAC	.25E-5	.20E-6	.57E-4	4.99
FMIN(1)	.11E-4	.87E-7	.74E-3	3.94
FMIN(2)	.86E-4	.62E-6	.40E-2	3.02
FMIN(3)	.11E-6	.57E-7	.55E-4	5.35
NLP	.11E-5	.19E-5	.12E-3	4.43
SUMT	.24E-5	.65E-10	.87E-3	4.00
DFP	.28E-6	.80E-8	.82E-4	5.42
FCDPAK	.27E-5	.49E-9	.22E-3	4.70

Table 2: Accuracy.

Code	ET	NF	NG	NDF	NDG
OPRQP	22.6	58	599	40	418
XROP	13.9	40	357	29	270
VF02AD	31.5	16	179	16	179
GRGA	37.7	204	2946	67	378
OPT	62.5	742	7528	0	321
GRG2(1)	52.6	297	3368	38	423
GRG2(2)	75.0	757	7677	0	0
VF01A	42.2	158	1595	158	603
LPNLP	57.4	252	2518	101	1014
SALQDR	41.6	120	1096	117	1068
SALQDF	88.2	1228	10731	0	0
SALMNF	132.0	78	4610	420	4610
CONMIN	99.9	955	9550	62	618
BIAS(1)	66.4	533	6621	65	729
BIAS(2)	148.1	1805	20901	0	0
FUNMIN	98.8	519	5023	112	1097
GAPFPR	73.1	147	1414	147	1414
GAPFQL	64.3	149	1439	149	1439
ACDPAC	70.5	222	4094	146	748
FMIN(1)	118.8	737	7027	158	1300
FMIN(2)	158.5	1168	18405	0	0
FMIN(3)	169.1	319	3242	620	6211
NLP	88.1	1043	8635	111	957
SUMT	270.1	2335	24046	99	1053
DFP	124.2	782	8214	107	1107
FCDPAK	23.1	125	480	63	251

Table 3a: Efficiency.

Code	ET	NF	NG	NDF	NDG
OPRQP	2	3	4	9	10
XROP	1	2	2	7	7
VF02AD	4	1	1	6	5
GRGA	5	10	10	13	9
OPT	10	18	18	1	8
GRG2(1)	8	13	12	8	11
GRG2(2)	15	19	19	1	1
VF01A	7	9	8	23	12
LPNLP	9	12	9	15	17
SALQDR	6	5	5	19	19
SALQDF	17	24	23	1	1
SALMNF	22	4	14	25	25
CONMIN	19	21	22	10	13
BIAS(1)	12	16	16	12	14
BIAS(2)	23	25	25	1	1
FUNMIN	18	15	15	18	20
GAPFPR	14	7	6	21	23
GAPFQL	11	8	7	22	24
ACDPAC	13	11	13	20	15
FMIN(1)	20	17	17	23	22
FMIN(2)	24	23	24	1	1
FMIN(3)	25	14	11	26	26
NLP	16	22	21	17	16
SUMT	26	26	26	14	18
DFP	21	20	20	16	21
FCDPAK	3	6	3	11	6

Table 3b: Efficiency (rank numbers).

Code	ET/A/B	NF/A/B	NG/A/B	NDF/A/B	NDG/A/B
OPRQP	1.5	4.0	4.1	35.1	376.4
XROP	1.0	3.1	2.4	27.9	244.0
VF02AD	1.9	1.0	1.0	13.6	142.4
GRGA	2.7	14.6	19.4	64.5	360.6
OPT	5.7	63.4	61.2	1.1	372.7
GRG2(1)	4.2	23.0	25.9	40.5	458.8
GRG2(2)	6.3	59.7	56.7	1.0	1.0
VF01A	2.5	9.1	8.7	123.1	459.7
LPNLP	5.1	21.5	20.2	121.3	1197.5
SALQDR	4.8	13.2	11.6	175.2	1603.8
SALQDF	10.2	140.0	115.5	1.4	1.4
SALMNF	10.0	6.2	30.9	418.0	4398.3
CONMIN	8.8	81.3	78.5	71.4	717.1
BIAS(1)	5.6	44.0	51.7	74.2	814.8
BIAS(2)	16.2	188.5	211.3	1.3	1.3
FUNMIN	5.7	28.8	26.3	84.8	821.8
GAPFPR	6.2	11.6	10.7	157.5	1519.1
GAPFQL	5.4	11.8	10.7	160.4	1527.7
ACDPAC	5.5	16.4	28.6	148.2	754.3
FMIN(1)	11.0	64.0	57.4	189.1	1532.8
FMIN(2)	17.1	237.9	182.8	1.4	1.4
FMIN(3)	13.1	22.8	23.0	627.1	6795.0
NLP	7.3	85.4	67.3	124.2	1059.0
SUMT	20.4	168.7	155.1	97.4	962.4
DFP	9.6	59.2	57.4	109.1	1091.1
FCDPAK	2.0	9.4	3.6	64.7	267.7

Table 4a: Efficiency related to accuracy.

Code	ET/A/B	NF/A/B	NG/A/B	NDF/A/B	NDG/A/B
OPRQP	2	3	4	8	10
XROP	1	2	2	7	6
VF02AD	3	1	1	6	5
GRGA	6	10	9	10	8
OPT	13	19	20	2	9
GRG2(1)	7	14	12	9	11
GRG2(2)	16	18	17	1	1
VF01A	5	5	5	18	12
LPNLP	9	12	10	17	20
SALQDR	8	9	8	23	24
SALQDF	21	23	23	4	4
SALMNF	20	4	15	25	25
CONMIN	18	21	22	12	13
BIAS(1)	12	16	16	13	15
BIAS(2)	24	25	26	3	2
FUNMIN	13	15	13	14	16
GAPFPR	15	7	6	22	21
GAPFQL	10	8	6	21	22
ACDPAC	11	11	14	20	14
FMIN(1)	22	20	18	24	23
FMIN(2)	25	26	25	4	3
FMIN(3)	23	13	11	26	26
NLP	17	22	21	19	18
SUMT	26	24	24	15	17
DFP	19	17	18	16	19
FCDPAK	4	6	3	11	7

Table 4b: Efficiency related to accuracy (rank numbers)

Code	PNS	FFV	FVC	F
OPRQP	23.2	.27E-7	.18E+1	1
XROP	31.2	.15E-6	.14E+2	0
VF02AD	6.2	.41E-1	.84E-4	5
GRGA	12.1	.12E-2	.22E-7	3
OPT	44.7	.41E-1	.76E-7	7
GRG2(1)	10.4	.99E-1	.35E-4	0
GRG2(2)	18.3	.53E-6	.75E-3	9
VF01A	23.9	.46E-6	.49E+1	13
LPNLP	26.7	.65E-7	.18E+3	10
SALQDR	50.7	.25E-3	.35E-4	13
SALQDF	48.7	.69E-3	.29E-4	15
SALMNF	15.6	.34E-5	.67E-3	18
CONMIN	57.6	.53E-4	.43E-3	14
BIAS(1)	25.0	.26E-4	.52E-6	4
BIAS(2)	30.8	.33E-4	.36E-5	13
FUNMIN	32.9	.25E-2	.98E-1	9
GAPFPR	21.4	.45E-2	.23E-7	24
GAPFQL	16.0	.14E-6	.49	30
ACDPAC	10.6	.96E-4	.74E-4	8
FMIN(1)	35.0	.35E-7	.16E-1	0
FMIN(2)	52.5	.52E-6	.85E-2	0
FMIN(3)	22.1	.19E-5	.87	15
NLP	15.6	.28E-7	.29E+1	31
SUMT	69.9	.11E-2	.40E-6	19
DFP	31.9	.27E+6	.80E+2	34
FCDPAK	23.9	.45E-4	.43E+1	2

Table 5a: Reliability.

Code	PNS	FFV	FVC	F
OPRQP	11	1	20	5
XROP	17	6	24	1
VFO2AD	1	23	11	9
GRGA	4	20	1	7
OPT	21	23	3	11
GRG2(1)	2	25	8	1
GRG2(2)	8	9	14	13
VFO1A	12	7	23	15
LPNLP	15	4	26	14
SALQDR	23	17	8	15
SALQDF	22	18	7	19
SALMNF	5	11	13	21
CONMIN	25	15	12	18
BIAS(1)	14	12	5	8
BIAS(2)	16	13	6	15
FUNMIN	19	21	17	9
GAPFPR	9	22	2	23
GAPFQL	7	5	18	24
ACDPAC	3	16	10	12
FMIN(1)	20	3	16	1
FMIN(2)	24	8	15	1
FMIN(3)	10	10	19	19
NLP	5	2	21	25
SUMT	26	19	4	22
DFP	18	26	25	26
FCDPAK	12	14	22	6

Table 5b: Reliability (rank numbers).

Code	PGS	GFV	GVC
OPRQP	19.2	-.46	.16E-7
XROP	17.6	-.37	.70E-8
VF02AD	25.6	-.11E+1	.23E-8
GRGA	28.6	-.10E+2	.60E-10
OPT	32.2	-.32E+1	.78E-7
GRG2(1)	20.0	-.26E+1	.16E-4
GRG2(2)	25.3	-.27E+1	.66E-5
VF01A	10.5	-.44	.12E-7
LPNLP	9.1	-.40E-1	.20E-5
SALQDR	33.3	-.11E+1	.58E-5
SALQDF	32.0	-.17E+1	.84E-5
SALMNF	24.2	-.12E+1	.17E-5
CONMIN	6.0	-.48	.15E-3
BIAS(1)	25.1	-.11E+2	.96E-5
BIAS(2)	17.3	-.10E+1	.96E-5
FUNMIN	17.9	-.19E+2	.76E-5
GAPFPR	21.2	-.63	.38E-6
GAPFQL	1.6	-.17E-1	.59E-9
ACDPAC	21.2	-.13E+1	.15E-5
FMIN(1)	16.0	-.58	.91E-5
FMIN(2)	10.5	-.13E+1	.15E-5
FMIN(3)	13.2	-.27E+1	.62E-5
NLP	18.5	-.16E+1	.19E-5
SUMT	16.4	-.19	.22E-4
DFP	26.6	-.19E+1	.35E-5
FCDPAK	28.7	-.94E+1	.56E-4

Table 6a: Global convergence.

Code	PGS	GFV	GVC
OPRQP	14	21	6
XROP	17	23	4
VF02AD	7	15	3
GRGA	5	3	1
OPT	2	5	7
GRG2(1)	13	8	23
GRG2(2)	8	6	17
VF01A	22	22	5
LPNLP	24	25	13
SALQDR	1	15	15
SALQDF	3	10	19
SALMNF	10	14	11
CONMIN	25	20	26
BIAS(1)	9	2	21
BIAS(2)	18	17	21
FUNMIN	16	1	18
GAPFPR	11	18	8
GAPFQL	26	26	2
ACDPAC	11	12	9
FMIN(1)	20	19	20
FMIN(2)	22	12	9
FMIN(3)	21	6	16
NLP	15	11	12
SUMT	19	24	24
DFP	6	9	14
FCDPAK	4	4	25

Table 6b: Global convergence (rank numbers).

Performance criterium	Notation	Items	Weights
Efficiency related to accuracy	E1	ET/A	.46
		NF/A	.07
		NG/A	.07
		NDF/A	.20
		NDG/A	.20
Efficiency related to accuracy	E2	ET/A	.06
		NF/A	.13
		NG/A	.13
		NDF/A	.34
		NDG/A	.34
Reliability	R	PNS	.51
		FFV	.11
		FVC	.19
		F	.19
Global convergence	G	PGS	.60
		GFV	.28
		GVC	.12

Table 7: Weights for evaluating performance criteria.

Code	E1	E2	R	G
OPRQP	5.01	7.15	10.47	15.00
XROP	3.34	5.00	13.97	17.12
VF02AD	3.72	4.18	6.84	8.76
GRGA	7.69	8.95	5.76	3.96
OPT	11.37	9.65	15.90	3.44
GRG2(1)	9.04	10.60	5.48	12.80
GRG2(2)	10.21	6.19	10.20	8.52
VF01A	9.00	11.80	14.11	19.96
LPNLP	13.08	15.98	15.69	22.96
SALQDR	14.27	18.67	17.97	6.60
SALQDF	14.68	10.30	18.14	6.88
SALMNF	20.53	20.67	10.22	11.24
CONMIN	16.29	15.17	20.10	23.72
BIAS(1)	13.36	14.40	10.93	8.48
BIAS(2)	15.61	9.77	13.58	18.08
FUNMIN	13.94	14.62	16.94	12.04
GAPFPR	16.21	16.87	11.76	12.60
GAPFQL	14.45	17.51	12.10	23.12
ACDPAC	13.61	15.47	7.47	11.04
FMIN(1)	22.25	22.37	13.76	19.72
FMIN(2)	16.47	10.51	16.16	17.64
FMIN(3)	22.66	22.18	13.42	16.20
NLP	18.23	19.19	11.51	13.52
SUMT	21.72	18.68	20.29	21.00
DFP	18.19	17.59	21.73	7.80
FCDPAK	6.07	7.53	12.98	6.52

Table 8: Weighted average rank numbers.

3. Performance for solving degenerate, ill-conditioned, and indefinite problems

When constructing degenerate, ill-conditioned, and indefinite
test problems as described in Chapter IV, it is not possible to
guarantee that the predetermined x* defines a global minimizer of
the underlying optimization problem. But in all numerical tests
we found no computed solution which could be accepted as an
approximation of a global minimizer with an objective function value
less than zero. We therefore assume that, verified by numerical
experiments, $\mathfrak{G} = \emptyset$ or, in other words, that x* is always the global
minimizer of the optimization problem. Nevertheless, we have to
decide if a current computed solution approximates x* or not and
we define the corresponding sets $\mathfrak{S}$ and $\mathfrak{R}$ in the following way:

$\overline{x} \in \mathfrak{S}$ if and only if $|f(\overline{x})| < \eta$ and $r(\overline{x}) < \eta$.

$\overline{x} \in \mathfrak{R}$ otherwise.
$\hfill (61)$

In our numerical tests, the tolerance η is set to

$$\eta := .01 \hfill (62)$$

Using the numerical data of Appendix A, we constructed 80 problems
in ten classes as described in Sections 4, 5, and 6 of Chapter IV.
The detailed results for final accuracy and efficiency of all classes
are evaluated for each code separately and are given in Appendix C.
To allow a condensed presentation and to give a review of the
performance of the optimization codes in these cases, we divide the
efficiency items ET/A, NF/A, NG/A, NDF/A, NDG/A, and the final accuracy A of two classes which represent the corresponding criterium in
an appropriate way. The accuracy A is obtained as the arithmetic
mean value of the numbers $a(\overline{x})$, cf. (59), for all successful solutions
$\overline{x}$ of the test problem class under consideration. The results are
shown in Tables 9, 10, and 11. The first column describes the optimization programs, the second one the change of the accuracy, and the
following ones the corresponding relative efficiency items execution
time and numbers of function and gradient evaluations. The abbreviations are RA, RET, RNF, RNG, RNDF, and RNDG, respectively.

Table 9 lists the performance results of all programs when solving degenerate problems. In this case, we divided the efficiency
and accuracy results obtained for the second class by those obtained
for the first class. The second class was chosen since it seems more
likely that the Lagrange multipliers of a practical optimization
problem differ in their order of magnitude than that they are zero.
In an analogous way, the performance of the optimization codes is
described in Tables 10 and 11 for the solution of ill-conditioned or
indefinite problems. In this case, we divided the results obtained
for the test problem classes 7 and 9 by those obtained for the
classes 5 and 8, respectively.

The Tables 9 to 11 consist of two parts. The first part, identified by 'a', contains the numerical data for RA, RET, RNF, RNG,
RNDF, and RNDG as described above. Furthermore one could try to get
a final score by evaluating average rank numbers. The individual
rank numbers are contained in part b of the tables. But since some
programs use numerical differentiation, each rank number rm of the
column headed by RNDF is replaced by $(5 \cdot rn - 1)/4$ and in the column
identified by RNDG the rank number rn is replaced by $(25 \cdot rn - 4)/21$.
As a consequence, the redefined rank numbers range between 1 and
26, the number of all competetive optimization programs. A final
score is then obtained by the average scaled rank numbers with
the weights defined by E1 in Table 7. In case of numerical differentiation, these weights are rescaled so that they sum up to unity.
The results are found in the columns headed by DE, IC, and ID,
respectively.

Code	RA	RET	RNF	RNG	RNDF	RNDG
OPRQP	1.06	.94	.99	.94	.94	.96
XROP	1.00	1.05	1.06	1.06	1.06	1.06
VF02AD	.97	1.01	1.04	1.04	1.04	1.04
GRGA	.97	1.32	1.31	1.65	1.08	1.08
OPT	1.14	1.14	1.17	1.08	–	1.27
GRG2(1)	.95	1.08	1.07	1.07	1.09	1.08
GRG2(2)	.94	.91	.91	.91	–	–
VF01A	1.04	.90	1.00	1.00	1.00	.75
LPNLP	1.25	.61	.63	.63	.61	.61
SALQDR	.97	1.51	1.49	1.49	1.56	1.56
SALQDF	1.00	.95	.95	.94	–	–
SALMNF	1.12	.83	.76	.82	.82	.82
CONMIN	1.01	1.01	1.08	1.08	1.04	1.04
BIAS(1)	1.24	.94	.96	.96	.92	.92
BIAS(2)	.98	1.22	1.23	1.23	–	–
FUNMIN	1.21	.90	.90	.90	.89	.89
GAPFPR	1.10	.62	.63	.63	.63	.63
GAPFQL	1.01	.92	.95	.95	.95	.95
ACDPAC	.95	2.16	2.25	2.31	2.37	2.08
FMIN(1)	1.07	.91	.91	.91	.91	.90
FMIN(2)	1.03	.86	.88	.88	–	–
FMIN(3)	1.02	.98	1.01	1.00	1.01	1.00
NLP	1.08	.71	.70	.70	.72	.72
SUMT	.98	1.01	1.04	1.04	1.04	1.04
DFP	.85	.89	.84	.84	.94	.94
FCDPAK	1.04	.80	.79	.73	.78	.74

<u>Table 9a</u>: Performance for solving degenerate problems.

Code	RET	RNF	RNG	RNDF	RNDG	DE
OPRQP	13	14	11	12.25	14.10	12.75
XROP	20	19	19	21.00	20.05	20.07
VF02AD	17	17	17	17.25	16.48	16.95
GRGA	24	24	25	22.25	21.24	23.17
OPT	22	22	21	-	23.62	22.32
GRG2(1)	21	20	20	23.50	21.24	21.41
GRG2(2)	10	9	9	-	-	9.76
VF01A	8	15	15	14.75	5.76	9.89
LPNLP	1	1	1	1.00	1.00	1.00
SALQDR	25	25	24	24.75	24.81	24.84
SALQDF	15	11	11	-	-	14.04
SALMNF	5	4	5	6.00	6.95	5.52
CONMIN	17	21	21	17.25	16.48	17.51
BIAS(1)	13	13	14	9.75	10.52	11.92
BIAS(2)	23	23	23	-	-	23.00
FUNMIN	8	8	8	7.25	8.14	7.88
GAPFPR	2	1	1	2.25	2.19	1.96
GAPFQL	12	11	13	13.50	12.90	12.48
ACDPAC	26	26	26	26.00	26.00	26.00
FMIN(1)	10	9	9	8.50	9.33	9.43
FMIN(2)	6	7	7	-	-	6.24
FMIN(3)	16	16	15	16.00	15.29	15.79
NLP	3	3	3	3.50	3.38	3.18
SUMT	17	17	17	17.25	16.48	16.94
DFP	7	6	6	11.00	11.71	8.60
FCDPAK	4	5	4	4.75	4.57	4.33

Table 9b: Performance for solving degenerate problems (rank numbers).

Code	RA	RET	RNF	RNG	RNDF	RNDG
OPRQP	1.14	2.05	2.72	2.72	1.92	1.92
XROP	.77	1.57	1.61	1.61	1.56	1.56
VFO2AD	.91	1.79	1.82	1.82	1.82	1.82
GRGA	.94	.69	.61	.75	.60	.67
OPT	.90	.47	.40	.48	–	.37
GRG2(1)	.74	1.90	2.10	2.09	1.62	1.62
GRG2(2)	.79	1.18	1.19	1.19	–	–
VFO1A	.46	6.63	7.17	7.17	7.17	6.44
LPNLP	.74	1.97	3.20	3.20	1.50	1.50
SALQDR	.75	1.70	1.76	1.76	1.69	1.69
SALQDF	.78	1.51	1.45	1.30	–	–
SALMNF	.74	1.17	1.65	1.16	1.16	1.16
CONMIN	.98	4.33	5.00	5.00	2.79	2.79
BIAS(1)	1.00	1.45	1.65	1.57	1.23	1.23
BIAS(2)	.80	1.68	1.70	1.70	–	–
FUNMIN	.71	2.83	2.86	2.86	2.84	2.84
GAPFPR	.95	.95	.93	.93	.93	.93
GAPFQL	.50	6.65	6.70	6.70	6.70	6.70
ACDPAC	1.01	1.52	1.58	1.56	1.58	1.50
FMIN(1)	.99	1.01	1.01	1.01	1.01	1.01
FMIN(2)	1.08	1.09	1.07	1.07	–	–
FMIN(3)	1.17	1.14	1.12	1.12	1.12	1.12
NLP	.94	1.43	1.30	1.30	1.63	1.63
SUMT	.90	1.12	1.10	1.10	1.17	1.17
DFP	.53	10.26	11.75	11.75	8.66	8.66
FCDPAK	.94	.86	.68	.81	.69	.82

<u>Table 10a</u>: Performance for solving ill-conditioned problems.

Code	RET	RNF	RNG	RNDF	RNDG	IC
OPRQP	21	20	20	19.75	20.05	20.42
XROP	15	13	15	12.25	14.10	14.13
VF02AD	18	18	18	18.50	18.86	18.27
GRGA	2	2	2	2.00	2.19	1.84
OPT	1	1	1	-	1.00	1.00
GRG2(1)	19	19	19	14.75	15.29	17.41
GRG2(2)	10	9	10	-	-	9.88
VF01A	24	25	25	24.75	23.62	24.21
LPNLP	20	22	22	11.00	11.71	16.82
SALQDR	17	17	17	17.25	17.67	17.19
SALQDF	13	11	11	-	-	12.52
SALMNF	9	14	9	7.25	8.14	8.83
CONMIN	23	23	23	21.00	21.24	22.25
BIAS(1)	12	14	14	9.75	10.52	11.53
BIAS(2)	16	16	16	-	-	16.00
FUNMIN	22	21	21	22.25	22.43	22.00
GAPFPR	4	4	4	3.50	4.57	4.01
GAPFQL	25	24	24	23.50	24.81	24.52
ACDPAC	14	12	13	13.50	11.71	13.23
FMIN(1)	5	5	5	4.75	5.76	5.10
FMIN(2)	6	6	6	-	-	6.00
FMIN(3)	8	8	8	6.00	6.95	7.39
NLP	11	10	11	16.00	16.48	13.03
SUMT	7	7	7	8.50	9.33	7.77
DFP	26	26	26	26.00	26.00	26.00
FCDPAK	3	3	3	2.25	3.38	2.93

Table 10b: Performance for solving ill-conditioned problems (rank numbers).

Code	RA	RET	RNF	RNG	RNDF	RNDG
OPRQP	1.03	1.03	1.04	1.04	1.02	1.02
XROP	.96	1.06	1.05	1.05	1.05	1.05
VF02AD	1.00	.94	.99	.99	.99	.99
GRGA	1.01	.98	.99	.99	1.00	.99
OPT	1.01	.94	.95	.94	-	.99
GRG2(1)	1.01	1.03	1.03	1.03	1.03	1.02
GRG2(2)	1.03	1.03	1.02	1.02	-	-
VF01A	1.00	.77	.78	.78	.78	.77
LPNLP	1.01	1.06	1.12	1.12	1.01	1.01
SALQDR	1.01	1.10	1.09	1.09	1.10	1.10
SALQDF	.89	1.15	1.11	1.10	-	-
SALMNF	.97	1.03	1.07	1.03	1.03	1.03
CONMIN	1.08	1.01	1.02	1.02	1.03	1.03
BIAS(1)	1.22	.81	.81	.81	.79	.79
BIAS(2)	1.17	.89	.88	.88	-	-
FUNMIN	.98	.99	.98	.98	.99	.99
GAPFPR	1.00	.91	.91	.91	.91	.91
GAPFQL	.81	2.91	2.90	2.90	2.90	2.90
ACDPAC	.88	1.00	1.05	1.03	1.01	.98
FMIN(1)	.91	1.06	1.07	1.07	1.08	1.08
FMIN(2)	1.00	1.00	1.00	1.00	-	-
FMIN(3)	1.02	.98	.96	.96	.98	.98
NLP	.98	.99	1.00	1.00	1.02	1.02
SUMT	1.00	1.02	1.02	1.02	.98	.98
DFP	1.04	1.07	1.07	1.07	1.04	1.04
FCDPAK	1.00	.98	.93	1.02	.93	1.04

Table 11a: Performance for solving indefinite problems.

Code	RET	RNF	RNG	RNDF	RNDG	ID
OPRQP	16	17	19	14.75	14.10	15.65
XROP	20	18	20	22.25	22.43	20.80
VF02AD	5	9	8	8.50	8.14	6.82
GRGA	7	9	8	11.00	8.14	8.24
OPT	5	6	5	-	8.14	5.88
GRG2(1)	16	16	16	17.25	14.10	15.87
GRG2(2)	16	13	12	-	-	15.18
VF01A	1	1	1	1.00	1.00	1.00
LPNLP	20	25	25	12.25	12.90	17.73
SALQDR	24	23	23	24.75	24.81	24.17
SALQDF	25	24	24	-	-	24.76
SALMNF	16	20	16	17.25	17.67	16.86
CONMIN	14	13	12	17.25	17.67	15.17
BIAS(1)	2	2	2	2.25	2.19	2.09
BIAS(2)	3	3	3	-	-	3.00
FUNMIN	10	8	7	8.50	8.14	8.98
GAPFPR	4	4	4	3.50	3.38	3.78
GAPFQL	26	26	26	26.00	26.00	26.00
ACDPAC	12	18	16	12.25	4.57	11.26
FMIN(1)	20	20	21	23.50	23.62	21.49
FMIN(2)	12	11	10	-	-	11.65
FMIN(3)	7	7	6	6.00	4.57	6.24
NLP	10	11	10	14.75	14.10	11.84
SUMT	15	13	12	6.00	4.57	10.76
DFP	23	20	21	21.00	20.05	21.66
FCDPAK	7	5	12	4.75	20.05	9.37

Table 11b: Performance for solving indefinite problems
(rank numbers)

4. Sensitivity to slight variations of the problem

The degenerate, ill-conditioned, and indefinite problems constructed for the tests of the previous section, possess a lot of
common properties. Specifically the problems of the first seven
classes possess the same feasible regions, and the active constraints
of the problems defining the classes 8, 9, and 10 are only augmented
by some linear functions. Nearly all information describing the
individual properties of degenerate, ill-conditioned, and indefinite
problems is contained in the quadratic terms of the objective function.

However, the numerical results of the last section or of Appendix
C show sometimes significant differences in the efficiency between
the classes and it could be useful to get some measure of whether one
optimization code is more sensitive to slight variations of the problem, especially of the objective function, than another. From this
one could conclude whether the performance of an algorithm is more
influenced by the special structure of the objective function or of
the constraints.

Therefore, we consider the final accuracy A and the efficiency
items ET/A, NF/A, NG/A, NDF/A, and NDG/A for each class separately
and evaluate the standard deviations over the ten classes for all
programs. Since the corresponding mean values differ widely in their
magnitude, we calculate the percentages of the deviations with
respect to the mean values. The results are presented in Table 12a
using the abbreviations SA, SET, SNF, SNG, SNDF, and SNDG, which
describe the sensitivity of a program with respect to accuracy,
execution time, and number of function and gradient evaluations.

To get a final score, we use the same strategy as in the previous section. First the modified rank numbers are evaluated, cf.
Table 12b, and subsequently the weighted average rank numbers are
computed for each program separately. The numerical results are
given in the column headed by SP.

Code	SA	SET	SNF	SNG	SNDF	SNDG
OPRQP	6.6	31.1	46.2	46.2	28.6	28.6
XROP	7.5	12.7	12.9	12.9	16.4	16.4
VFO2AD	4.7	30.2	21.0	21.0	21.0	21.0
GRGA	12.5	64.4	61.3	96.1	41.5	40.0
OPT	14.1	38.8	43.3	39.9	-	41.3
GRG2(1)	14.9	35.0	51.6	51.6	19.0	19.0
GRG2(2)	10.0	42.4	25.5	25.5	-	-
VFO1A	19.0	88.0	96.1	96.1	96.1	86.1
LPNLP	39.7	79.8	87.9	87.9	77.5	77.5
SALQDR	17.2	32.6	35.4	35.4	33.4	33.4
SALQDF	25.7	26.6	26.7	27.7	-	-
SALMNF	46.9	58.6	68.2	58.7	58.7	58.7
CONMIN	13.7	89.5	100.5	100.5	67.7	67.7
BIAS(1)	26.8	64.1	62.7	63.0	67.3	67.3
BIAS(2)	18.0	48.5	49.0	48.9	-	-
FUNMIN	15.8	70.5	72.1	72.1	70.1	70.1
GAPFPR	6.0	25.2	24.1	24.1	24.1	24.1
GAPFQL	24.1	67.9	69.5	69.5	69.5	69.5
ACDPAC	13.0	35.9	37.2	37.4	37.8	36.9
FMIN(1)	3.6	4.7	3.9	3.9	4.1	4.1
FMIN(2)	11.7	28.2	29.3	29.3	-	-
FMIN(3)	13.4	16.8	16.1	16.1	18.4	18.4
NLP	12.3	26.1	24.7	24.7	29.0	29.0
SUMT	8.0	8.8	11.1	8.9	12.0	10.1
DFP	35.6	112.5	124.9	124.9	104.9	104.9
FCDPAK	5.0	27.4	45.2	31.1	40.8	28.8

Table 12a: Sensitivity to slight variations of the problem.

Code	SET	SNF	SNG	SNDF	SNDG	SP
OPRQP	11	15	15	9.75	9.33	10.98
XROP	3	3	3	3.50	3.38	3.18
VF02AD	10	5	5	7.25	6.95	8.14
GRGA	20	18	23	16.00	15.29	18.33
OPT	15	13	14	-	16.48	15.10
GRG2(1)	13	17	17	6.00	5.76	10.71
GRG2(2)	16	8	8	-	-	14.13
VF01A	24	24	23	24.75	24.81	24.24
LPNLP	23	23	22	23.50	23.62	23.15
SALQDR	12	11	12	12.25	12.90	12.16
SALQDF	7	9	9	-	-	7.48
SALMNF	18	20	18	17.25	17.67	17.92
CONMIN	25	25	25	19.75	20.05	22.96
BIAS(1)	19	19	19	18.50	18.86	18.87
BIAS(2)	17	16	16	-	-	16.76
FUNMIN	22	22	21	22.25	22.43	22.07
GAPFPR	5	6	6	8.50	8.14	6.47
GAPFQL	21	21	20	21.00	21.24	20.98
ACDPAC	14	12	13	13.50	14.10	13.71
FMIN(1)	1	1	1	1.00	1.00	1.00
FMIN(2)	9	10	10	-	-	9.24
FMIN(3)	4	4	4	4.75	4.57	4.26
NLP	6	7	7	11.00	11.71	8.28
SUMT	2	2	2	2.25	2.19	2.09
DFP	26	26	26	26.00	26.00	26.00
FCDPAK	8	14	11	14.75	10.52	10.48

Table 12b: Sensitivity to slight variations of the problem (rank numbers).

5. Sensitivity to the position of the starting point

As described in Section 7 of Chapter IV, we constructed 25 convex test problems combined in 5 different classes. The corresponding data are presented in Appendix A. Each problem is provided with a starting point close to the solution and another one far away from the solution to test the sensitivity of an optimization program to the position of the starting point. Since the predetermined solution x* is a global one for convex problems, we have $\mathfrak{G} = \emptyset$. For the same reason, we may define a relatively high tolerance

$$\eta := .1 \qquad\qquad (63)$$

for determining the set $\mathfrak{S}$ of successful solutions and the set $\mathfrak{N}$ of non-successful solutions using the definition (61) of Section 3.

To evaluate the sensitivity to the position of the starting point, we are interested in the relative efficiency of the optimization codes. We therefore calculate the accuracy A and the efficiency items ET/A, NF/A, NG/A, NDF/A, and NDG/A, first for all problems with a starting point close to the solution and subsequently for all problems with a starting point far away from the solution. Then we divide the last results by those of the first test series to obtain the relative accuracy RA and the relative efficiency items RET, RNF, RNG, RNDF, and RNDG. If a test run finished with a failure or a non-successful solution, we suppress the corresponding result for the other starting point when evaluating the mean values. The results are presented in Table 13a.

Table 13b shows the modified rank numbers and, in particular, the weighted average rank numbers in the column defined by SS. These data are determined in the same way as in the previous sections. More detailed information and the original efficiency results are given in Appendix C for both test series.

But the test results presented so far do not give an answer to the question, if the sensitivity is a consequence of an excellent local convergence rate or of a worse global convergence rate when starting far away from the solution. For a more thorough investigation of these questions, one could divide the efficiency scores

NF/A and NDF/A of both test series by the corresponding results obtained for the general test problem category as described in Section 2 of this chapter. The resulting scores are denoted by RNFL, RNDFL for starting points close to the solution and RNFG, RNDFG for starting points far away from the solution, and are presented in Table 13c. The analogous evaluations for relative execution times and numbers of restriction calls are omitted, since convex exponential sums are computed much faster than signomials, and since the test problems are formulated with different numbers of restrictions. A more condensed impression of the performance can be obtained by evaluating average weighted rank numbers as in the previous sections. The column headed by SC in Table 13c gives the rank numbers for the local convergence rate of the algorithms and SF denotes the rank numbers for the global convergence rate.

Code	RA	RET	RNF	RNG	RNDF	RNDG
OPRQP	.99	2.38	2.36	2.72	1.86	2.04
XROP	1.05	3.11	2.62	3.29	1.99	3.35
VFO2AD	1.02	2.39	2.30	2.43	2.30	2.43
GRGA	1.37	4.72	2.00	6.07	3.04	3.29
OPT	.98	3.20	2.18	3.18	–	2.69
GRG2(1)	1.12	1.74	2.27	2.43	1.31	1.37
GRG2(2)	1.10	.99	1.25	.96	–	–
VFO1A	.98	2.80	1.82	1.79	1.82	2.75
LPNLP	1.01	1.02	.99	1.00	1.06	1.10
SALQDR	1.23	1.81	1.49	1.54	1.51	1.56
SALQDF	1.06	1.58	1.50	1.57	–	–
SALMNF	.99	1.81	2.04	1.81	1.66	1.81
CONMIN	.70	1.64	1.69	1.33	1.47	1.18
BIAS(1)	.94	2.65	3.40	2.89	2.14	2.12
BIAS(2)	1.22	1.55	1.73	1.56	–	–
FUNMIN	.76	8.12	7.98	7.68	7.85	7.56
GAPFPR	1.08	.65	.82	.72	.82	.71
GAPFQL	1.00	4.24	2.06	2.58	2.06	2.58
ACDPAC	.98	1.61	1.73	1.66	1.57	1.46
FMIN(1)	.82	1.31	1.86	1.95	1.11	.73
FMIN(2)	.70	1.45	1.58	1.36	–	–
FMIN(3)	.71	.88	1.45	1.56	.98	.73
NLP	1.01	1.43	1.28	1.45	1.39	1.62
SUMT	.97	1.30	1.34	1.30	1.34	1.34
DFP	.95	2.77	2.43	2.69	2.46	2.82
FCDPAK	1.21	9.82	11.00	13.09	8.41	10.08

Table 13a: Sensitivity to the position of the starting point.

Code	RET	RNF	RNG	RNDF	RNDG	SS
OPRQP	16	21	20	16.00	14.10	16.25
XROP	21	23	23	17.25	23.62	21.05
VFO2AD	17	20	16	21.00	16.48	17.84
GRGA	24	15	24	23.50	22.43	22.96
OPT	22	18	22	-	18.86	20.86
GRG2(1)	13	19	16	6.00	8.14	11.26
GRG2(2)	3	3	2	-	-	2.88
VFO1A	20	13	13	14.75	20.05	17.98
LPNLP	4	2	3	3.50	4.57	3.80
SALQDR	14	7	8	11.00	10.52	11.79
SALQDF	10	8	11	-	-	9.88
SALMNF	14	16	14	13.50	12.90	13.82
CONMIN	12	10	5	9.75	5.76	9.67
BIAS(1)	18	24	21	19.75	15.29	18.44
BIAS(2)	9	11	9	-	-	9.24
FUNMIN	25	25	25	24.75	24.81	24.91
GAPFPR	1	1	1	1.00	1.00	1.00
GAPFQL	23	17	18	18.50	17.67	20.26
ACDPAC	11	11	12	12.25	9.33	10.99
FMIN(1)	6	14	15	4.75	2.19	6.18
FMIN(2)	8	9	6	-	-	7.88
FMIN(3)	2	6	9	2.25	2.19	2.86
NLP	7	4	7	8.50	11.71	8.03
SUMT	5	5	4	7.25	6.95	5.77
DFP	19	22	19	22.25	21.24	20.31
FCDPAK	26	26	26	26.00	26.00	26.00

Table 13b: Sensitivity to the position of the starting point
             (rank numbers).

Code	RNFL	RNDFL	SC	RNFG	RNDFG	SF
OPRQP	.35	.52	7	.84	.96	4
XROP	.30	.42	3	.79	.84	3
VF02AD	.72	.72	14	1.65	1.65	18
GRGA	.51	.65	12	1.98	1.96	21
OPT	.79	-	16	1.81	-	18
GRG2(1)	1.41	1.67	25	3.65	2.43	24
GRG2(2)	2.86	-	26	3.58	-	25
VF01A	.55	.55	8	1.00	1.00	6
LPNLP	.57	.40	5	.57	.42	1
SALQDR	.70	.71	13	1.12	1.14	11
SALQDF	.72	-	15	1.09	-	10
SALMNF	.26	.33	1	.53	.54	2
CONMIN	.92	1.12	20	1.55	1.65	17
BIAS(1)	.59	.64	11	2.00	1.35	20
BIAS(2)	1.19	-	21	1.50	-	15
FUNMIN	.40	.43	4	3.23	3.41	26
GAPFPR	.97	.97	19	1.06	1.06	8
GAPFQL	.56	.56	9	1.14	1.14	12
ACDPAC	.56	.64	10	.97	1.00	7
FMIN(1)	1.12	1.16	22	1.97	1.34	16
FMIN(2)	1.37	-	23	1.28	-	13
FMIN(3)	.76	.76	17	1.07	.71	4
NLP	2.27	1.54	24	2.89	2.16	22
SUMT	.85	.88	18	1.15	1.19	14
DFP	.41	.44	6	1.01	1.10	9
FCDPAK	.29	.34	2	3.21	2.88	23

Table 13c: Local and global convergence rate.

6. Ease of use

For the practical usage of optimization programs and especially for their implementation in a user oriented program library, it is most important to get programs which could be handled in a simple, non-sophisticated way. This requires to develop some kind of measure for the performance criterium 'ease of use'. From our point of view, we describe this criterium by the following four items:

a) <u>Quality of documentation</u>: We can observe significant differences in the quality of documentation according as a code is part of a program library, for example, or is considered as an experimental, preliminary version.

b) <u>Provision of problem data and functions</u>: All programs under consideration require different strategies to provide the code with the problem data like dimension, number of constraints, stopping parameters, etc., and the corresponding problem functions f, $g_1,...,g_m$ together with their gradients. In some cases, these informations can be supplied in a simple, straightforward way, but in other cases the user has to do a lot of additional work to prepare the problem, for example to transform inequality into equality constraints, upper and lower bounds into general restrictions, etc.. All these additional implementations raise the probability to incorporate programming errors and make it more difficult to solve problems.

c) <u>Program organization</u>: The decision, if a program is better organized than another one, is mainly influenced by the length of the code and the possibility to use variable dimensions, numerical differentiations, and a flag to allow a return to the driving program whenever a user wants to do it.

d) <u>Sensitivity to input parameters</u>: Most programs require the definition of some special input parameters, for example stopping tolerances, initial penalty parameter, reduction factor, and the decision between different stopping strategies, steplength procedures, update algorithms, etc.. Whenever possible, we used default values or those which were recommended by the authors. But we could not follow their recommendations in all cases, and when trying to find

a suitable combination of all possible parameters, we got an impression
if a code is more sensible to slight variations of the input parameters
than another one.

First we are concerned with the difficulty that it is not possible
to evaluate the four items in any directly measurable way. To over-
come this difficulty we apply Saaty's [SY 1975, SY 1977] priority
theory which proceeds from considering each pair of programs separately and to decide, if there are more or less significant differences
in the four performance items defined above. A rough sketch of the
theory and the detailed numerical analysis are outlined in Appendix D.
Table 15a contains the scores of each item using the following abbreviations:

QD - Quality of documentation.

PD - Provision of problem data and functions.

PO - Program organization.

SI - Sensitivity to input parameters.

A higher score has to be interpreted in the sense that the corresponding program is prefered to another one with a lower score. As in the
previous sections, we are interested now in getting a measure for the
underlying performance criterium, in this case of 'ease of use'. First
we have to provide the four items with weight factors which are to be
chosen by a decision maker. As an example, we present an evaluation
based on the weights of Table 14 which are obtained by a multicriteria
analysis, too, cf. Appendix D. In case of disagreement, a reader
should define his own weight factors. As a final score for the ease of
use of optimization programs, we evaluate the weighted average rank
numbers as in the previous sections. The individual rank numbers and
the final scores denoted by EU are presented in Table 15b.

Item	Weight
QD	.36
PD	.23
PO	.13
SI	.28

Table 14: Weights for evaluating ease of use.

Code	QD	PD	PO	SI
OPRQP XROP	.043	.054	.020	.077
VF02AD	.043	.087	.142	.095
GRGA	.036	.138	.022	.101
OPT	.079	.032	.062	.033
GRG2(1/2)	.076	.062	.038	.071
VF01A	.044	.066	.062	.067
LPNLP	.105	.053	.038	.033
SALQDR SALQDF SALMNF	.115	.147	.142	.066
CONMIN	.029	.011	.034	.014
BIAS(1/2)	.075	.032	.092	.033
FUNMIN	.014	.032	.034	.022
GAPFPR	.016	.024	.020	.033
GAPFQL	.020	.115	.027	.059
ACDPAC	.075	.032	.027	.101
FMIN(1/2/3)	.015	.039	.092	.016
NLP	.079	.022	.059	.046
SUMT	.029	.015	.038	.011
DFP	.033	.016	.022	.020
FCDPAK	.075	.022	.027	.101

Table 15a: Ease of use.

Code	QD	PD	PO	SI	EU
OPRQP XROP }	14	10	24	5	11.86
VFO2AD	14	6	1	4	7.67
GRGA	17	4	22	1	10.18
OPT	5	16	10	15	10.98
GRG2(1/2)	7	8	13	7	8.01
VFO1A	13	7	10	9	10.11
LPNLP	4	12	13	15	10.09
SALQDR SALQDF } SALMNF	1	1	1	10	3.52
CONMIN	19	26	17	25	22.03
BIAS(1/2)	9	16	5	15	11.77
FUNMIN	26	16	17	20	20.85
GAPFPR	22	21	24	15	20.07
GAPFQL	21	5	19	13	14.82
ACDPAC	9	16	19	1	9.67
FMIN(1/2/3)	23	13	5	22	18.08
NLP	5	22	12	14	12.34
SUMT	19	25	13	26	21.56
DFP	18	24	22	21	20.74
FCDPAK	9	22	19	1	11.05

Table 15b: Ease of use (rank numbers).

7. How to get a final score

A decision maker might now be interested in selecting an optimization program which will solve his problems in the most appropriate way. First, the program may have to satisfy special technical or organisatorial constraints, e.g., embedded numerical differentiation. This information is obtained by investigating the data of Table 1 and defines a special subset of programs for a final decision. This decision has to be based on the numerical results of the previous sections, but the performance of the optimization programs is evaluated by nine different criteria. It is therefore necessary to define weight factors describing the relative significance of the performance criteria. They depend on the individual usage of the programs and a test designer cannot, of course, prescribe the weights, but he should point out possibilities of obtaining and applying significance factors for getting a final score.

One possible way to obtain weight factors for the performance criteria is described in the form of a multicriteria analysis in Appendix D, which includes three examples. From these examples, we got the weights of Table 18, however, we would like to stress that a reader should define his own weights in case of disagreement. By the proposed significance factors, we try to realize situations where efficiency, ease of use, or reliability, respectively, are the most important features of an optimization program.

The next step consists of considering the scores which are obtained for the performance criteria separately. The are summarized in Table 16a, where the efficiency results corresponding to E1 are presented, cf. Table 8. Since all these results are evaluated in form of weighted average mean values and each criterium consists of three to five items, these data again depend on the individual significance defined by the decision maker and they have to be replaced by analogous ones if he intends to use his own weights. By multiplying these scores with the corresponding weights of Table 18 and adding the products, we get the final scores for three different combinations of the weights as shown in Table 17a. Table 16b contains the rank numbers of all performance criteria and Table 17b the corresponding rank numbers of the final scores.

Code	E	R	G	DE	IC	ID	SP	SS	EU
OPRQP	5.0	10.5	15.0	12.8	20.4	15.7	11.0	16.3	11.9
XROP	3.3	14.0	17.1	20.1	14.1	20.8	3.2	21.1	11.9
VFO2AD	3.7	6.8	8.8	17.0	18.3	6.8	8.1	17.8	7.7
GRGA	7.7	5.8	4.0	23.2	1.8	8.2	18.3	23.0	10.2
OPT	11.4	15.9	3.4	22.3	1.0	5.9	15.1	20.9	11.0
GRG2(1)	9.0	5.5	12.8	21.4	17.4	15.9	10.7	11.3	8.0
GRG2(2)	10.2	10.2	8.5	9.8	9.9	15.2	14.1	2.9	8.0
VFO1A	9.0	14.1	20.0	9.9	24.2	1.0	24.2	18.0	10.1
LPNLP	13.1	15.7	23.0	1.0	16.8	17.7	23.2	3.8	10.1
SALQDR	14.3	18.0	6.6	24.8	17.2	24.2	12.2	11.8	3.5
SALQDF	14.7	18.1	6.9	14.0	12.5	24.8	7.5	9.9	3.5
SALMNF	20.5	10.2	11.2	5.5	8.8	16.9	17.9	13.8	3.5
CONMIN	16.3	20.1	23.7	17.5	22.3	15.2	23.0	9.7	22.0
BIAS(1)	13.4	10.9	8.5	11.9	11.5	2.1	18.9	18.4	10.8
BIAS(2)	15.6	13.6	18.1	23.0	16.0	3.0	16.8	9.2	10.8
FUNMIN	13.9	16.9	12.0	7.9	22.0	9.0	22.1	24.9	20.9
GAPFPR	16.2	11.8	12.6	2.0	4.0	3.8	6.5	1.0	20.1
GAPFQL	14.5	12.1	23.1	12.5	24.5	26.0	21.0	20.3	14.8
ACDPAC	13.6	7.5	11.0	26.0	13.2	11.3	13.7	11.0	9.7
FMIN(1)	22.3	13.8	19.7	9.4	5.1	21.5	1.0	6.2	18.1
FMIN(2)	16.5	16.2	17.6	6.2	6.0	11.7	9.2	7.9	18.1
FMIN(3)	22.7	13.4	16.2	15.8	7.4	6.2	4.3	2.9	18.1
NLP	18.2	11.5	13.5	3.2	13.0	11.8	8.3	8.0	12.3
SUMT	21.7	20.3	21.0	16.9	7.8	10.8	2.1	5.8	21.6
DFP	18.2	21.7	7.8	8.6	26.0	21.7	26.0	20.3	20.7
FCDPAK	6.1	13.0	6.5	4.3	2.9	9.4	10.5	26.0	11.1

Table 16a: Summarized scores for all performance criteria.

Code	E	R	G	DE	IC	ID	SP	SS	EU
OPRQP	3	7	16	14	21	17	12	16	15
XROP	1	16	18	20	15	21	3	23	15
VFO2AD	2	3	9	17	20	7	7	17	4
GRGA	5	2	2	24	2	8	19	24	10
OPT	9	19	1	22	1	5	16	22	11
GRG2(1)	7	1	14	21	19	18	11	13	5
GRG2(2)	8	5	8	10	10	16	15	3	5
VFO1A	6	17	22	11	24	1	25	18	9
LPNLP	10	18	24	1	17	20	24	4	8
SALQDR	14	22	4	25	18	24	13	14	1
SALQDF	16	23	5	15	12	25	6	11	1
SALMNF	23	6	11	5	9	19	18	15	1
CONMIN	19	24	26	19	23	16	23	10	26
BIAS(1)	11	8	7	12	11	2	20	19	13
BIAS(2)	17	14	20	23	16	3	17	9	13
FUNMIN	13	21	12	7	22	9	22	25	24
GAPFPR	18	10	13	2	4	4	5	1	22
GAPFQL	15	11	25	14	25	26	21	20	18
ACDPAC	12	4	10	26	14	12	14	12	7
FMIN(1)	25	15	21	9	5	22	1	6	19
FMIN(2)	20	20	19	6	6	13	9	7	19
FMIN(3)	26	13	17	16	7	6	4	2	19
NLP	22	9	15	3	13	14	8	8	17
SUMT	24	25	23	17	8	11	2	5	25
DFP	21	26	6	8	26	23	26	21	23
FCDPAK	4	12	3	4	3	10	10	26	12

Table 16b: Summarized scores for all performance criteria (rank numbers).

Code	Q1	Q2	Q3
OPRQP	10.47	11.52	11.64
XROP	11.21	12.07	13.09
VFO2AD	8.00	8.41	8.25
GRGA	9.18	9.28	8.33
OPT	12.36	11.64	12.03
GRG2(1)	9.82	9.60	9.48
GRG2(2)	9.48	9.10	9.59
VFO1A	12.86	12.88	14.83
LPNLP	13.43	13.04	15.66
SALQDR	13.74	11.14	13.53
SALQDF	12.90	10.42	12.86
SALMNF	13.04	10.23	11.96
CONMIN	18.63	19.74	19.85
BIAS(1)	12.21	11.95	11.57
BIAS(2)	14.40	13.62	14.57
FUNMIN	16.41	17.80	16.32
GAPFPR	12.39	13.73	11.57
GAPFQL	16.01	16.29	16.70
ACDPAC	11.79	11.04	10.67
FMIN(1)	16.23	16.02	15.15
FMIN(2)	14.72	15.26	15.02
FMIN(3)	15.84	15.52	14.18
NLP	13.27	12.72	12.33
SUMT	18.34	18.49	18.02
DFP	18.93	19.53	18.26
FCDPAK	9.78	10.28	10.51

Table 17a: Final scores.

Code	Q1	Q2	Q3
OPRQP	6	10	9
XROP	7	13	14
VFO2AD	1	1	1
GRGA	2	3	2
OPT	10	11	11
GRG2(1)	5	4	3
GRG2(2)	3	2	4
VFO1A	12	15	18
LPNLP	16	16	21
SALQDR	17	9	15
SALQDF	13	7	13
SALMNF	14	5	10
CONMIN	25	26	26
BIAS(1)	9	12	7
BIAS(2)	18	17	17
FUNMIN	23	23	22
GAPFPR	11	18	8
GAPFQL	21	22	23
ACDPAC	8	8	6
FMIN(1)	22	21	20
FMIN(2)	19	19	19
FMIN(3)	20	20	16
NLP	15	14	12
SUMT	24	24	24
DFP	26	25	25
FCDPAK	4	6	5

Table 17b: Final scores (rank numbers).

Performance criterium		Q1	Q2	Q3
Efficiency	E	.32	.18	.14
Reliability	R	.23	.18	.36
Global convergence	G	.08	.08	.20
Performance for solving degenerate problems	DE	.05	.03	.03
Performance for solving ill-conditioned problems	IC	.05	.06	.03
Performance for solving indefinite problems	ID	.03	.03	.03
Sensitivity to variations of the problem	SP	.03	.03	.06
Sensitivity to the position of the starting point	SS	.07	.06	.06
Ease of use	EU	.14	.35	.09

Table 18: Weights for evaluating final scores.

Chapter VI ....

CONCLUSIONS, RECOMMENDATIONS, REMARKS

Based on the numerical results of the previous chapter, we first
present some conclusions and try to explain extraordinary phenomena.
Subsequently we give recommendations for the design of optimization
programs which could be useful for future software developments and
numerical experiments. The last section contains some remarks
and technical details about this comparative study of optimization
programs.

1. Final conclusions

Before drawing any conclusions, the reader should be aware of some
facts influencing the numerical results. The first major difficulty
arises when determining the stopping tolerances, maximum iteration
numbers, first guesses for a penalty parameter, and so on, which are
required to execute an optimization program. As far as possible, we
used those data which are predetermined or recommended by the authors.
But we could not follow their recommendations in all cases, especially
since very often they were not related to the underlying machine precision. It was a difficult task to find an appropriate combination of
all possible parameters and we cannot give any guarantee for having
found the best combination. Nevertheless we feel that an unjustified
determination of a parameter does not alter the overall performance
significantly in all cases. For example, a smaller stopping tolerance
leads to higher calculation times and more function or gradient
evaluations, but, on the other hand, to an increased accuracy. A
small maximum iteration number or any similar criterion leads to a
decreased reliability on the one hand, but to an increased efficiency
on the other and vice versa. A further difficulty arises when

comparing the efficiency of a reliable code with that of an unreliable code, since we have to expect that the latter could not solve the difficult, time-consuming problems. If in such a case two programs possess about the same efficiency, one should accept the more reliable one as the more efficient program. Furthermore we have to note that unreliability could lead to slightly improved global convergence, since any termination point with negative objective function value is counted as a global solution although we do not know if the necessary conditions are satisfied. Finally, the tolerances $\eta_1, \ldots, \eta_4$ influence the numerical results since the number of successful, global, or non-successful solutions is closely related to these parameters. Therefore we have to relate the percentages of global or non-successful solutions to the corresponding objective function values and constraint violations, since smaller tolerances, for example, increase the percentage of global solutions, but decrease the absolute cost function values, and vice versa.

All the difficulties mentioned above could influence the measurable results and, in addition, the non-measurable ones defining ease of use depend on the subjective impression of the test designer as to whether one factor is more or less significant than another. A decision maker should interprete the results carefully when selecting a code. In the following, we give some general conclusions about the performance of optimization programs.

Efficiency: First investigation of the results shows the remarkable progress which has been made in recent years to improve the efficiency of optimization programs. Especially the more drastical reduction of function or gradient evaluations compared with the reduction of calculation time indicates that modern optimization codes require an extensive amount of internal operations before computing a new function value. A rough ranking of the underlying mathematical methods indicates that the quadratic approximation methods and the method of Robinson are the most efficient, followed by the generalized reduced gradient, multiplier, and penalty methods. However, the most surprising result is obtained by an extremely low number of function evaluations of VF02AD, and we expect that further investigations of this promising algorithm will reduce calculation time too. The introduction of numerical differentiation in some programs increases execution time and objective and constraint function calls, but the final efficiency score is improved, if a decision maker attaches a high significance to the number of gradient evaluations.

Reliability: The results of Tables 5 and 8 indicate that reliability depends mainly on the individual implementation of an algorithm and not on the underlying mathematical method. It is necessary to relate the percentage of non-successful solutions to the corresponding objective function values and constraint violations. Although VFO2AD shows the lowest percentage of non-successful solutions, the program has higher scores for FFV and FVC than GRGA, for example, because this program yields some non-successful solutions which are only poor approximations of the predetermined minimizer. The generalized reduced gradient methods retain the feasibility of the current iterates to a high accuracy, since these methods try to follow the constraint boundaries as far as possible. Therefore, we recommend generalized reduced gradient algorithms for all applications where feasible solutions are most desirable. Most failures are caused by overflow and excessive calculation time, but also unexpectedly often by zero division.

Global convergence: As above, we observe that global convergence depends mainly on the individual implementation of an algorithm, in particular on the line-search procedure used, with one exception: The generalized reduced gradient methods possess the best global convergence behavior. Since we should distinguish between non-successful approximations of the predetermined minimizer on the one hand and approximations of global minimizers with negative objective function values on the other, we have to relate the percentage of global solutions to the obtained objective function values. When evaluating the global convergence results as proposed in the previous chapter, we proceeded from the assumption that a high percentage of global solutions is most desirable. In most cases, a user is indeed more interested in getting a global than a local solution. Nevertheless there are applications where a local solution closest to the starting point has to be computed. In these cases, one should define the corresponding rank numbers in the opposite way.

Performance for solving degenerate problems: The quadratic approximation methods OPRQP, XROP, and VFO2AD do not change their efficiency significantly when solving degenerate or nearly degenerate problems.

In particular, they do not take any advantage of redundant constraints.
On the other hand, the generalized reduced gradient methods GRGA and
OPT have some numerical difficulties in the sense that efficiency
items, in particular execution time and number of constraint function
evaluations, are correlated to the degree of degeneracy. One possible
explanation for the bad efficiency in this case could be that the
evaluation of the reduced gradient leads to numerical difficulties
since the norm of the gradient of the objective function and its
angle with the tangent plane of the constraints decrease when ap-
proaching the optimal solution. Most multiplier methods show a slight
improvement of efficiency when solving nearly degenerate problems
and a drastical improvement when all Lagrange multipliers vanish
especially if we relate efficiency to the achieved accuracy. This
indicates that multiplier methods are often able to handle problems
of this kind as unconstrained optimization problems. Exceptional
cases are CONMIN without any significant change of efficiency and
ACDPAC and SALQDR which even show reduced efficiency when solving
nearly degenerate problems. The penalty methods FMIN(1/2/3) and SUMT
do not change their efficiency significantly in contrast to NLP and
DFP which show a performance similar to multiplier methods.

Performance for solving ill-conditioned problems: Almost all programs
needed higher calculation times and much more function or gradient
evaluations when solving ill-conditioned problems. The most drastic
deterioration was observed for VFO1A, GAPFQL, and DFP. The last program
seems to be unable to enter the acceleration phase for ill-conditined
problems. FMIN(1/2/3) and SUMT show no significant change of their
performance, but the most remarkable results are obtained by the
generalized reduced gradient methods GRGA, OPT and by Robinson's
method FCDPAK which could even improve their efficiency when solving
ill-conditioned problems. Since these methods try to follow the
constraint boundaries as far as possible, they are able to prevent
the iterates from entering the deep, extended valleys describing
the curvature of ill-conditioned Lagrangians in the neighbourhood of
an optimal solution. The improvement of the efficiency could be explained by the 'compression' of the curvature of the Lagrangian
when changing the matrix P as described in Section 5, because this
effect leads to a steeper curvature of the Lagrangian along the
constraint surface.

<u>Performance for solving indefinite problems</u>: Besides GAPFQL, the
programs had no exceptional difficulties when solving indefinite
problems. At most slightly higher calculation times or numbers
of function and gradient evaluations were observed. The multiplier
methods VFO1A and BIAS(1/2) even showed improved efficiency for
the solution of indefinite problems.

<u>Sensitivity to slight variations of the problem</u>: Some programs, e.g.,
DFP, VFO1A, CONMIN, LPNLP, and FUNMIN seem to be very sensitive
to slight variations of the test problems and, in particular, of
the objective functions in contrast to the penalty methods FMIN(1)
and SUMT which produced about the same efficiency measures for all
test problem classes. For an explanation, we have to consider the
underlying mathematical method. FMIN(1) and SUMT are interior penalty
methods in the sense that at least some of the constraints are used
to define logarithmic barrier terms. These terms together with the
penalty parameter destroy the curvature of f inside the feasible
region so that even a drastical change of f does not influence the
penalty function remarkably. Only the introduction of numerical
differentiation lead to a higher sensitivity.

<u>Sensitivity to the position of the starting point</u>: All programs are
more or less sensitive to the position of the starting point, with
three exceptions: LPNLP shows nearly no change of efficiency,
FMIN(1/2/3) needs a lower number of iteration steps when starting
far away from the solution, and GAPFPR solves these problems even
more efficiently. The three programs are able to take advantage of
the convex structure of a problem. On the other hand, the multiplier
method FUNMIN, the generalized reduced gradient method GRGA, and
Robinson's method FCDPAK are the most sensitive ones.

<u>Ease of use</u>: All optimization codes which are incorporated in a pro-
gram library show the best scores for ease of use in contrast to
the older ones like CONMIN and SUMT. This is a consequence of the
wide distribution of the library programs for practical application
in the hands of non-specialized users.

2. Recommendations for the design of optimization programs

Based on extensive experiences with the implementation and test
of optimization programs, we give some recommendations for the design
and organization of nonlinear programming codes. They should be considered as an assistance for future developments from the viewpoint
of a user. Of course, we cannot give any recommendations about the
mathematical structure of the algorithm and we restrict ourself only
to some technical implementation details which could be realized in
every program independently of the underlying algorithm. We proceed
from the situation that a designer is developing a code with the
intention to distribute it to non-specialized users and to execute
it on different computers.

Problem formulation: When looking for an adequate problem formulation,
we recommend format (1) since most optimization programs of this
study proceed from this formulation. Nevertheless, any equivalent
formulation could be used.

Programming language: All programs which are submitted for the comparative study are written in FORTRAN. It is therefore advisable
to use the same language for future algorithms, in particular,
if the new code has to replace an existing one. Whenever possible,
the designer should use a portable version of FORTRAN, for example
ANSI FORTRAN. Furthermore, the program should not contain any auxiliary subroutines in machine language.

Precision: Since the machine precision varies drastically from one
computer to another, an optimization program should be available in
single and double precision versions.

Documentation: A clearly written and detailed documentation facili-
tates the solution of problems for all users who are not familar with
the program. As a model, we recommend the qualified documentation of
the programs contained in the Optimization Software Library of the
National Physical Laboratory. A documentation of this library consists
of the following sections:

Purpose
Description
Specification
Parameters
User-supplied subroutines
Error indicators
Auxiliary routines
Storage
Timing
Accuracy
Further comments
Keywords
References.

In addition, there is implementation information and a detailed
example.

Provision of problem data: In general, it is possible to supply the
problem data such as dimension, number of constraints, tolerances,
etc., within a user-written driving program or to read them in on an
input file. The latter should be preferred if unsophisticated users
have to solve individual problems, but there could be some organi-
sational difficulties when a program has to be integrated into a
complicated system. If a designer is in doubt, he should write the
program as a subroutine and supply the problem data within the main
program in common areas or subroutine arguments. But the best re-
commendation would be to offer both versions.

Provision of problem functions: How an optimization program could
be provided with the problem functions depends on the underlying
mathematical algorithm. But a designer should try to require the
evaluation of as many problem functions and gradients as possible
in one subroutine or block.

Numerical differentiation: It should be self-evident that a user-
oriented optimization program possesses the capability to calculate
derivatives numerically.

Linear constraints: The solution of a problem is facilitated if the
user may supply linear constraints and bounds of the variables

separately and not in the form of general equality or inequality restrictions. In particular, this reduces the probability of implemen-
ting programming errors.

Parameters: An optimization program requires definition of parameters
for stopping criteria, steplength procedures, maximum number of
iterations, etc.. A designer should predetermine them by default
values depending on the underlying machine precision, but a user
should be informed about possible alterations and their effects on
performance, if the default values do not lead to a successful solution.

Flag: The flexibility of an optimization program is increased if a
user may leave the subroutine calculating the problem functions, and
return to the main program whenever he wants. The implementation of
a flag is therefore advised.

Output: The output should consist of multiple print levels to follow
the course of computations as closely as necessary. If the solution
of a problem could not be terminated successfully, the user should
get complete error information. On the other hand, the user should
have the possibility to suppress all output.

Dimensioning: An optimization program should be variably dimensioned
so that the required storage capacity is adapted to the problem size.
As an example for variably dimensioned arrays, consider the FORTRAN
subroutines in Lawson, Hanson [LH 1974].

3. Some technical details

This section contains some additional information about performance evaluation and some technical details on the realization of
this comparative study.

Computer: All numerical tests are performed on a Telefunken TR440 computer at the Rechenzentrum of the University of Würzburg using the standard FORTRAN compiler PS&FTNCOMP of Telefunken. The calculations within the driving programs are carried out in single precision with more than 10 correct digits (35 - 38 bit mantissa).

Random number generator: The random number generator is written in Telefunken Assembler (TAS) and produces a sequence of uniformly distributed numbers in the interval (0,1) in the following way, cf. Coveyou and MacPherson [CM 1967]:

$$x_0 = 1$$
$$x_{n+1} = (ax_n + b) \text{ modulo } p \, , \quad n=0,1,\dots$$
$$a = 31415 \; 92653 \; 5897$$
$$b = 14868 \; 96974 \; 5365$$
$$p = 2^{46}$$

(64)

Classical test problems: In spite of the disadvantages mentioned earlier, we have collected a large number of test problems from the literature which have been used in the past to test and compare optimization software. More than 120 problems are documented and compiled together with their derivatives, cf. Hock and Schittkowski [HS 1980]. All programs will also be executed on these problems. The corresponding tests are performed by W. Hock and the results will be published elsewhere.

Earlier study: To compare the numerical results with those of an earlier study, the reader is referred to the Thesis of Sandgren [SA 1977], where 17 programs in 35 different versions are compared using 30 real life test problems and numerical differentiation. Most of the codes are realizations of penalty methods, with the exception of four generalized reduced gradient and two linear approximation algorithms. The codes tested in both studies are OPT, GRG2, GREG (an older version of GRGA), BIAS, FMIN, and SUMT. Although Sandgren presented a different evaluation system, we found nearly the same ranking of the programs with one exception: In our study, GRGA performed better than OPT. This could be a consequence of the improvements of GREG and the fact that OPT does not allow for providing the program with analytical derivatives of the objective function.

Submission of programs: All programs were obtained from their authors in the original form on magnetic tapes, card decks, or paper tapes. Some technical modifications were often necessary since most programs are not written in a completely portable FORTRAN language. Furthermore we had to transform double precision programs into single precision arithmetic, and we had to prevent the problem data being read in with the card reader unit number. It was not possible to implement all optimization codes which were submitted for the comparative study. Some of the underlying mathematical algorithms seemed to be unable to solve our highly nonlinear test problems, and some codes had to be rejected because of unreliable programming. We did not try to get all existing penalty methods, since they are now superseded by multiplier, generalized reduced gradient, and quadratic approximation methods.

Capacity: An attentive reader will miss any information on the capacity of optimization programs, i.e., the storage requirements related to the size of the problem. But it is not possible to find an appropriate measure for capacity since many programs have fixed array dimensions leading to a lot of unused storage locations.

Accuracy: The results of Table 2 show that the final accuracy differs widely from one code to another. It is not practicable to require that all programs yield the same predetermined final accuracy for two reasons: First, the optimization programs use completely different stopping criteria with varying effects on the final accuracies in objective function value, constraint violations, Kuhn-Tucker condition, and number of exact digits. The second reason is that some codes could not achieve a higher accuracy because of excessive calculation times or internal reasons.

Extent: The optimization programs tested for this comparative study, consist of several hundreds of subroutines with a total length of about 64,000 FORTRAN statements. In addition, we had to implement the test problem generator, the classical test problems, the driving programs, and some programs to evaluate the results with an extent of about 25,000 statements. More than half a million bytes were required to store the data describing the randomly generated test problems and the test results. About 380 hours CPU-time were spent to perform the tests.

Appendix A
. . . .

NUMERICAL DATA FOR CONSTRUCTING TEST PROBLEMS

The numerical tests of an optimization program are based on three
sets of test problems. The first set is designed to test the overall
performance of a code, i.e. efficiency, reliability, and global convergence. To generate the test problems as described in Section 2 of
Chapter IV, the user has to set the following parameters:

n :   Dimension of the problem.

m :   Number of constraints.

m_e :   Number of equality constraints.

m_a :   Number of active inequality constraints.

xl,xu :   Lower and upper bounds for the coefficients of the
                optimal solution x^*.

k :   Number of terms in each signomial s_j, $j=0,1,\ldots,m$.

cl_1,cu_1 :   Lower and upper bounds for the coefficients c_i of s_o.

i_1 :   Declares if the exponents of s_o are integers (= T) or
                not (= F).

p_1 :   Percentage of zero exponents in each term of s_o.

al_1,au_1 :   Lower and upper bounds for the exponents a_{ij} of s_o.

cl_2,cu_2 :   Lower and upper bounds for the coefficients of s_1,
                $l=1,\ldots,m$.

i_2 :   Declares if the exponents of s_1, $l=1,\ldots,m$, are integers
                (= T) or not (= F).

p_2 :   Percentage of zero exponents in each term of s_1, $l=1,\ldots,m$.

al_2,au_2 :   Lower and upper bounds for the exponents a_{ij} of each
                s_1, $l=1,\ldots,m$.

b_1,b_2 :   Lower and upper bounds for the Lagrange multipliers
                u_j^*, $j=1,\ldots,m_e$.

b_3 :   Upper bound for the Lagrange multipliers u_j^*, $j=m_e+1,\ldots,m_e+m_a$.

b_4,b_5 :   Lower and upper bounds for the elements of the upper triangular matrix U.

By choosing any numerical values for the above parameters, the user defines one special class of test problems and a repeated execution of the random number generator gives arbitrarily many different problems of the class. For our numerical tests we used the data of Table 19 to determine ten classes of test problems.

Class	1A	2A	3A	4A	5A	6A	7A	8A	9A	10A
n	4	5	6	7	8	9	10	12	15	20
m	6	12	3	10	10	8	13	14	5	18
m_e	1	0	3	0	0	0	2	0	2	0
m_a	2	5	0	3	4	3	5	8	1	12
xl	1	3	3.5	8	4	2	0.5	1.5	5.2	0.4
xu	4	4	4.5	10	5	3	1	1.6	5.8	0.6
k	5	7	6	8	4	2	6	4	4	5
cl_1	-1	-10	-1	-1	0	0	-2	5	2	-1
cu_1	1	10	0	20	1	5	50	20	15	10
i_1	F	F	T	T	T	F	T	T	F	T
p_1	50	70	30	70	20	60	30	5	10	90
al_1	1	-4	-5	-4	-3	-2	2	-3	-4	1
au_1	3	4	5	0	1	0	6	1	-1	5
cl_2	-5	0	-2	-1	1	-6	-8	-3	-10	-9
cu_2	5	5	2	4	5	10	1	15	0	1
i_2	T	T	F	T	F	T	T	T	F	T
p_2	50	70	30	50	40	75	10	5	20	95
al_2	-2	-5	-1	-5	-2	-3	0	-2	-4	0
au_2	3	5	1	0	-1	1	5	1	2	4
b_1	-1	-1	-1	-1	-1	-1	-1	-1	-1	-1
b_2	1	1	1	1	1	1	1	1	1	1
b_3	1	1	1	1	1	1	1	1	1	1
b_4	-1	-1	-1	-1	-1	-1	-1	-1	-1	-1
b_5	1	1	1	1	1	1	1	1	1	1

<u>Table 19</u>: Data for ten test problem classes (general case).

Using a random number generator with a uniform distribution, cf. (64), we computed ten test problems for the classes 1A,...,6A, and five test problems for the classes 7A,...,10A. Each of the 80 problems is provided with three different starting points whose coefficients are randomly chosen between the upper and lower bounds

$$x_1 := (x1,\ldots,x1)^T \quad, \quad x_u := (xu,\ldots,xu)^T$$

leading to 240 test runs of an optimization code under consideration. To give an example, we describe the first test problem of the class 1A. All real numbers are rounded to make the problem data readable. Of course, these data are stored in full accuracy for their numerical execution. The signomials $s_0,\ldots,s_m$ are the following ones:

$$s_0(x) = .32x_1^{2.82}x_2^{2.16} - .88x_3^{2.86} + .81x_2^{1.07}x_3^{1.48}x_4^{1.40}$$
$$+ .66x_1^{1.56} - .83x_1^{2.30}x_2^{2.40}x_3^{2.27}$$

$$s_1(x) = .96 - 1.12x_3x_4^2 + 3.13x_1^{-1}x_2^{-1}x_3^{-1}x_4^2 + 3.30x_1^2$$
$$- 4.84x_1^{-1}x_3$$

$$s_2(x) = 5.39 + 1.98x_2x_3^2x_4^{-1} - 4.14x_2 - 1.88x_1x_4$$

$$s_3(x) = .99 - 4.64x_3x_4^{-1} - .57x_4$$

$$s_4(x) = -3.23 - 1.89x_2^2 - 1.18x_3^2 - 2.01x_2^2x_4^{-1} - 1.99x_1$$

$$s_5(x) = 3.45 - 2.98x_1x_2 - .96x_1^2x_3^2 - 2.90x_2^{-1}x_4^{-1} + 3.39x_3^2$$

$$s_6(x) = -.51 - 2.89x_2^2x_3 - 6.24x_2^2 + 2.15x_3^2x_4 \; .$$

The optimal solution x* is given by

$$x^* = (1.35 \; , \; 2.17 \; , \; 2.80 \; , \; 3.71)^T \; .$$

Using x* and the formulae (43), it is possible to determine the restrictions $g_1,\ldots,g_m$ leading to the system

$$g_1(x) = s_1(x) + 40.99 = 0$$
$$g_2(x) = s_2(x) + \; 3.98 \geq 0$$
$$g_3(x) = s_3(x) + \; 4.61 \geq 0$$
$$g_4(x) = s_4(x) + 32.06 \geq 0$$

$$g_5(x) = s_5(x) - 2.71 \geq 0$$
$$g_6(x) = s_6(x) + 7.27 \geq 0 .$$

The second and third restrictions are active at x*. Following the instructions of Section 1 of Chapter IV, the data determining the quadratic term of f, namely H, q, and α, are given by

$$H = \begin{pmatrix} 171.49 & 200.07 & 156.82 & -3.12 \\ 200.07 & 78.48 & 86.63 & -12.85 \\ 156.82 & 86.63 & 51.75 & -20.46 \\ -3.12 & -12.85 & -20.46 & -5.07 \end{pmatrix} ,$$

$$q = (-908.19 , -546.53 , -407.35 , 64.17)^T ,$$

$$\alpha = 1499.97 .$$

The second test problem set was designed to check the performance of optimization codes in extreme situations. Therefore we determined ten classes of degenerate, ill-conditioned, and indefinite test problems as described in Sections 4, 5, and 6 of Chapter IV. To allow comparisons of the results as far as possible, the problems of two classes differ at most in the linear terms of the restrictions d_j, $j=1,\ldots,m_e+m_a$, the optimal Lagrange multipliers u_j^*, $j=1,\ldots,m_e+m_a$, or in the Hessian of the Lagrangian P. All remaining data are fixed in the following way:

$$n = 8$$
$$m = 10$$
$$m_e = 0$$
$$m_a = 4$$
$$xl = 4 , xu = 5$$
$$cl_1 = 0 , cu_1 = 1$$
$$i_1 = T$$
$$p_1 = 20$$
$$al_1 = -3 , au_1 = 1$$
$$cl_2 = 1 , cu_2 = 5$$
$$i_2 = F$$
$$p_2 = 40$$

$$al_2 = -2 \ , \ au_2 = -1 \ .$$

Every test problem class contains 8 problems. The random number generator was restarted for each class, so that we always got the same series of signomials and the same optimal solutions. Since we are only interested in the local convergence behavior of the optimization programs, we define rather close bounds of the kind

$$x_1 := x^* - e \ , \ x_u := x^* + e \ , \ e := (.1,\ldots,.1)^T \ .$$

In addition, each test problem is supplied with a randomly chosen starting point within the bounds x_1 and x_u.

To obtain degenerate test problems, we let $d_j = 0$, $j=1,\ldots,m_a$, and the elements of the upper triangular matrix U are randomly chosen between -1 and 1. As proposed in Section 4 of Chapter IV, we generated four classes of test problems distinguished by their optimal Lagrange multipliers, cf. Table 20.

Class	1B	2B	3B	4B
u_1^*	1	1	1	0
u_2^*	1	10^{-2}	1	0
u_3^*	1	10^{-4}	0	0
u_4^*	1	10^{-6}	0	0

Table 20: Lagrange multipliers for degenerate test problems.

Ill-conditioned test problems are only distinguished by the Hessian of the Lagrangian or, more precisely, by the values of ν, cf. (55). The linear terms are set to $d_j = 0$, $j=1,\ldots,m_e+m_a$, and the Lagrange multipliers are randomly chosen in the interval $(0,1)$. The three values of ν determining the classes 5B, 6B, and 7B are set out in Table 21.

Class	5B	6B	7B
ν	3	5	8

Table 21: Values of ν for determining ill-conditioned problems.

Following the guidelines of Section 6, Chapter IV, it is easy to
compute indefinite test problems. The Lagrange multipliers are randomly chosen in the interval (0,1) and the elements of the upper triangular matrix U are randomly chosen in (-1,1). The parameter σ controls
whether the Hessian of the Lagrangian is indefinite or not and determines the last three test problem classes, see Table 22.

Class	8B	9B	10B
σ	.1	-.1	0

Table 22: Values of σ for determining
indefinite test problems.

The third test problem set contains convex optimization problems
as described in Section 7 of Chapter IV to check the sensitivity of
a program to the position of the starting point. In this case, the
signomials are replaced by exponential sums of the kind (53), but we
can construct them by an analogous set of parameters with only one
exception: For convex problems, we omit the upper triangular matrix
U and therefore the bounds b_4 and b_5. The remaining parameters are
listed in Table 23 to define five classes of convex test problems.
For each class, we generated five optimization problems of type
(1) randomly and all problems were provided with two different starting points which are selected in the following way: First, we determine
a vector $\bar{x}$ randomly with coefficients between x_1 and x_u. Then we define values

$$x_{01} := x^* + .1(\bar{x} - x^*)$$
$$x_{02} := x^* + 10(\bar{x} - x^*)$$

which are used as starting points for the two test series. The upper
and lower bounds are adapted in an analogous way. Numerical experiments show that x^* always defines a unique minimizer so that the
addition of a positive definite matrix P' to P was not necessary.

Class	1C	2C	3C	4C	5C
n	4	6	8	10	15
m	6	3	10	13	5
m_e	1	3	0	2	2
m_a	2	0	4	5	1
xl	-1	-1	-1	-1	-1
xu	1	1	1	1	1
k	5	6	4	6	4
cl_1	0	0	0	0	0
cu_1	5	6	4	6	4
i_1	F	T	T	T	F
p_1	50	30	20	30	10
al_1	.1	-.5	-.3	.2	-.4
au_1	.3	.5	.1	.6	-.1
cl_2	-5	-6	-4	-6	-4
cu_2	0	0	0	0	0
i_2	T	F	F	T	F
p_2	50	30	40	10	20
al_2	-.2	-.1	-.2	0	-.4
au_2	.3	.1	-.1	.5	.2
b_1	0	0	0	0	0
b_2	1	1	1	1	1
b_3	1	1	1	1	1

Table 23: Data for five test problem classes (convex case).

Appendix B
....

SENSITIVITY ANALYSIS FOR THE TEST PROBLEMS

For reasons pointed out in Section 1, Chapter IV, we have to check for each problem whether the predetermined and therefore exact minimizer x* defines a solution of the computed optimization problem numerically. It is not sufficient to calculate objective function value or sum of constraint violations at x* only, since the definition of $f(x^*)$ or $g_j(x^*)$, $j=1,\ldots,m$, uses differences and all digits could be extinguished although any small perturbation of x* could give inadequate high values. Therefore we propose a sensitivity analysis in the following way: The solution x* is perturbed by

$$x_\epsilon^* := x^* + \epsilon e$$

with $e := (1,-1,\ldots,(-1)^{n-1})^T$ and small tolerances ϵ in accordance with the machine precision. For these perturbed solutions, objective function values, sums of constraint violations, and the Euclidean norms of the Kuhn-Tucker vectors were evaluated. To summarize the results, we computed the corresponding geometric mean values for each of the 25 test problem classes using the tolerances

$$\epsilon = 0 \ , \ \epsilon = 10^{-9} \ , \ \epsilon = 10^{-7} \ .$$

The abbreviations in Tables 24 to 26 are:

FV_ϵ : Geometric mean of the absolute objective function values $f(x_\epsilon^*)$ for one test problem class.

VC_ϵ : Geometric mean values of the sums of constraint violations $r(x_\epsilon^*)$ for one test problem class.

KT_ϵ : Geometric mean values of the Euclidean norms of the Kuhn-Tucker vectors $h(x_\epsilon^*)$ for one test problem class.

Class	ϵ	FV_ϵ	VC_ϵ	KT_ϵ
	0	0	0	.62E-22
1A	E-9	.29E-7	.64E-7	.27E-8
	E-7	.29E-5	.64E-5	.27E-6
	0	0	0	.75E-21
2A	E-9	.38E-5	.13E-4	.37E-8
	E-7	.38E-3	.14E-2	.37E-6
	0	.21E-11	.95E-11	.62E-10
3A	E-9	.79E-9	.29E-8	.98E-8
	E-7	.80E-7	.30 E-6	.99E-6
	0	.14E-12	.10E-11	.55E-12
4A	E-9	.42E-11	.10E-10	.58E-8
	E-7	.26E-9	.99E-9	.58E-6
	0	0	0	.19E-20
5A	E-9	.49E-8	.88E-8	.47E-8
	E-7	.49E-6	.88E-6	.11E-5
	0	.37E-11	.14E-11	.12E-11
6A	E-9	.67E-11	.11E-10	.58E-8
	E-7	.33E-9	.11E-8	.59E-6
	0	.57E-11	.42E-10	.15E-10
7A	E-9	.25E-9	.14E-8	.73E-8
	E-7	.22E-7	.14E-6	.73E-6
	0	.17E-10	.77E-10	.11E-9
8A	E-9	.21E-8	.25E-7	.13E-7
	E-7	.21E-6	.25E-5	.13E-5
	0	0	.10E-15	.32E-13
9A	E-9	.18E-9	.49E-9	.16E-7
	E-7	.19E-7	.50E-7	.16E-5
	0	.16E-10	.19E-9	.17E-9
10A	E-9	.49E-8	.69E-7	.29E-7
	E-7	.49E-6	.69E-5	.29E-5

Table 24: Sensitivity analysis for ten test problem
classes (general problems).

Class	ϵ	FV_ϵ	VC_ϵ	KT_ϵ
	O	.19E-12	.12E-11	.10E-11
1B	E-9	.45E-11	.14E-10	.31E-8
	E-7	.53E-9	.13E-8	.31E-6
	O	.19E-12	.12E-11	.69E-12
2B	E-9	.46E-11	.14E-10	.31E-8
	E-7	.31E-9	.13E-8	.31E-6
	O	.19E-12	.12E-11	.10E-11
3B	E-9	.47E-11	.14E-10	.31E-8
	E-7	.42E-9	.13E-8	.31E-6
	O	.19E-12	.12E-11	.11E-12
4B	E-9	.64E-13	.14E-10	.31E-8
	E-7	.11E-12	.13E-8	.31E-6
	O	.19E-12	.12E-11	.77E-12
5B	E-9	.67E-11	.14E-10	.27E-8
	E-7	.36E-9	.13E-8	.27E-6
	O	.19E-12	.12E-11	.77E-12
6B	E-9	.67E-11	.14E-10	.22E-8
	E-7	.36E-9	.13E-8	.22E-6
	O	.19E-12	.12E-11	.77E-12
7B	E-9	.67E-11	.14E-10	.79E-9
	E-7	.36E-9	.13E-8	.79E-7
	O	.19E-12	.12E-11	.77E-12
8B	E-9	.20E-9	.44E-8	.10E-8
	E-7	.20E-7	.44E-6	.10E-6
	O	.19E-12	.12E-11	.77E-12
9B	E-9	.20E-9	.44E-8	.10E-7
	E-7	.20E-7	.44E-6	.10E-6
	O	.19E-12	.12E-11	.77E-12
10B	E-9	.20E-9	.44E-8	.99E-9
	E-7	.20E-7	.44E-6	.99E-7

Table 25: Sensitivity analysis for ten test problem
classes (degenerate, ill-conditioned, and
indefinite problems).

Class	ε	FV_ε	VC_ε	KT_ε
	0	0	0	0
1C	E-9	.93E-9	.40E-8	.67E-9
	E-7	.87E-7	.39E-6	.69E-7
	0	0	0	0
2C	E-9	.11E-8	.29E-8	.35E-8
	E-7	.12E-6	.27E-6	.36E-6
	0	0	0	0
3C	E-9	.99E-9	.24E-8	.49E-9
	E-7	.97E-7	.23E-6	.49E-7
	0	0	0	.11E-23
4C	E-9	.28E-8	.16E-7	.53E-8
	E-7	.28E-6	.16E-5	.54E-6
	0	0	0	.84E-24
5C	E-9	.33E-8	.13E-7	.65E-8
	E-7	.34E-6	.13E-5	.65E-6

Table 26: Sensitivity analysis for five test problem classes (convex problems).

FURTHER RESULTS

The numerical results describing efficiency, reliability, and global convergence of an optimization program depend on the tolerances $\eta_1, \ldots, \eta_4$ on which the decision between successful, non-successful, and global solutions depends. These tolerances could influence the final performance evaluation, especially for codes with a lower accuracy. We now replace the tolerances (60) by the more restrictive ones

$$\eta_1 := .0001 \ , \ \eta_2 := .0001 \ , \ \eta_3 := .01 \ , \ \eta_4 := 3$$

and present the analogous performance results in Tables 27 to 32. For evaluating the average rank numbers of Table 32, we used the weights of Table 7.

Although stronger tolerances $\eta_1, \ldots, \eta_4$ imply a higher final accuracy, most optimization programs were able to improve their efficiency. Obviously, some of the more difficult, time-consuming problems could be solved only with a lower accuracy and are now omitted in the evaluation of new average mean values of the results of the successful test runs. The most drastic changes in the performance are observed for codes which could not achieve a high accuracy, since a lot of non-successful test runs are now only poor approximations of the predetermined minimizer. Nevertheless, the evaluation of average rank numbers takes this into account so that these rank numbers differ significantly from the earlier ones only in very few cases.

Subsequently, we report the individual performance results for each optimization program separately. More precisely, the average accuracy, efficiency, reliability, and global convergence items are computed for the test problem classes 1A,...,10A, and we evaluate the average accuracy and efficiency items for the classes 1B,...,10B. This allows for the study of performance for separate classes, for example for problems with only equality constraints, with many active con-

straints, or zero multipliers. To evaluate the results for the first
ten classes 1A,...,10A, we proceeded from the original guess of the
tolerances $\eta_1,...,\eta_4$, cf. (60). The last row of the tables describing
the performance for degenerate, ill-conditioned, and indefinite problems contains the number of non-successful solutions.

We finish this chapter by presenting more detailed results obtained
for the solution of convex test problems. Table 85 contains the final
accuracy A and the efficiency items ET, NF, NG, NDF, NDG for starting
points close to the solution. The column headed by R gives the number
of non-successful test runs. The corresponding average performance
results for all test runs with starting points far away from the solution are reported in Table 86.

The usage of exponential sums for the definition of test problems
lead to a lot of runs which were interrupted because of overflow.
These failures occured more often during the execution of the higher
dimensional, time-consuming test problems, so that the overall calculation time is relatively low for programs with a great number of
failures. We point out that the average mean values of Tables 85 and
86 are obtained only from those test runs which were successful for
both starting points. In most cases, the calculation times are lower
for both test series compared with the calculation times of the
general purpose tests, since the evaluation of an exponential sum is
much faster than the evaluation of a signomial.

Code	FV	VC	KT	ED
OPRQP	.55E-8	.43E-8	.19E-6	7.68
XROP	.24E-7	.11E-7	.38E-5	6.60
VF02AD	.22E-8	.31E-10	.26E-5	6.77
GRGA	.59E-6	.59E-11	.56E-4	5.66
OPT	.36E-5	.12E-8	.75E-3	4.49
GRG2(1)	.20E-6	.21E-7	.21E-4	5.99
GRG2(2)	.69E-6	.64E-8	.61E-3	4.63
VF01A	.15E-8	.86E-9	.29E-6	7.67
LPNLP	.31E-6	.49E-6	.13E-4	5.61
SALQDR	.21E-5	.46E-6	.39E-3	4.43
SALQDF	.19E-5	.13E-6	.35E-3	4.37
SALMNF	.31E-7	.48E-7	.72E-5	5.99
CONMIN	.46E-7	.95E-8	.29E-4	5.84
BIAS(1)	.11E-5	.14E-5	.60E-4	4.91
BIAS(2)	.28E-5	.24E-5	.45E-3	4.12
FUNMIN	.38E-9	.33E-9	.64E-5	6.33
GAPFPR	.13E-5	.34E-10	.64E-3	4.40
GAPFQL	.55E-6	.10E-5	.68E-5	5.53
ACDPAC	.21E-5	.17E-6	.48E-4	5.15
FMIN(1)	.17E-5	.34E-7	.17E-3	4.59
FMIN(2)	.19E-4	.19E-4	.30E-2	3.47
FMIN(3)	.37E-7	.22E-7	.24E-4	5.69
NLP	.84E-6	.19E-5	.97E-4	4.58
SUMT	.40E-6	.11E-10	.37E-3	4.47
DFP	.50E-8	.26E-9	.31E-4	5.89
FCDPAK	.14E-5	.23E-9	.17E-3	4.83

Table 27: Accuracy.

Code	ET	NF	NG	NDF	NDG
OPRQP	19.5	49	491	36	367
XROP	12.4	34	326	27	268
VF02AD	29.9	15	164	15	164
GRGA	33.4	191	2570	65	327
OPT	54.5	722	7685	0	277
GRG2(1)	49.4	284	3023	37	387
GRG2(2)	70.4	737	7914	0	0
VF01A	40.1	148	1507	148	581
LPNLP	51.3	228	2368	97	995
SALQDR	36.3	156	1063	153	1044
SALQDF	60.1	837	6487	0	0
SALMNF	120.3	72	4282	388	4282
CONMIN	81.5	798	7851	53	525
BIAS(1)	61.8	486	5850	59	642
BIAS(2)	102.7	1344	14608	0	0
FUNMIN	99.7	521	5102	112	1114
GAPFPR	46.6	113	1005	113	1005
GAPFQL	53.0	130	1289	130	1289
ACDPAC	68.4	218	4192	145	771
FMIN(1)	139.5	599	5525	239	2088
FMIN(2)	129.5	2047	15530	0	0
FMIN(3)	160.1	322	3067	578	5050
NLP	81.7	990	8410	108	950
SUMT	292.8	2439	27383	107	1231
DFP	105.3	639	7034	93	1012
FCDPAK	20.4	128	451	64	235

Table 28a: Efficiency.

Code	ET	NF	NG	NDF	NDG
OPRQP	2	3	4	8	10
XROP	1	2	2	7	7
VF02AD	4	1	1	6	5
GRGA	5	10	10	13	9
OPT	12	19	20	1	8
GRG2(1)	9	13	11	9	11
GRG2(2)	16	20	22	1	9
VF01A	7	8	8	22	13
LPNLP	10	12	9	15	17
SALQDR	6	9	6	23	20
SALQDF	13	22	18	1	1
SALMNF	22	4	14	25	25
CONMIN	17	21	21	10	12
BIAS(1)	14	15	17	11	14
BIAS(2)	20	24	24	1	1
FUNMIN	19	16	15	18	21
GAPFPR	8	5	5	19	18
GAPFQL	11	7	7	20	23
ACDPAC	15	11	13	21	15
FMIN(1)	24	17	16	24	24
FMIN(2)	23	25	25	0	0
FMIN(3)	25	14	12	26	26
NLP	18	23	23	17	16
SUMT	26	26	26	16	22
DFP	21	18	19	14	19
FCDPAK	3	6	3	12	6

Table 28b: Efficiency (rank numbers).

Code	ET/A/B	NF/A/B	NG/A/B	NDF/A/B	NDG/A/B
OPRQP	1.5	3.3	3.2	28.7	286.2
XROP	1.0	2.5	2.3	23.6	227.0
VFO2AD	2.3	1.0	1.0	12.3	124.6
GRGA	2.7	13.6	16.9	56.7	280.3
OPT	6.0	65.0	68.5	1.0	308.0
GRG2(1)	4.7	22.9	24.8	37.7	393.9
GRG2(2)	7.3	63.3	67.1	1.0	1.0
VFO1A	2.8	8.7	8.7	106.8	424.4
LPNLP	5.5	18.9	19.5	106.9	1094.6
SALQDR	4.1	14.8	10.1	179.5	1237.2
SALQDF	6.5	75.2	59.9	1.1	1.1
SALMNF	10.5	6.0	28.9	353.6	3604.6
CONMIN	7.3	61.1	60.1	50.7	500.8
BIAS(1)	6.6	42.8	51.0	65.4	702.0
BIAS(2)	12.6	137.6	151.2	1.2	1.2
FUNMIN	7.3	31.6	30.8	84.5	839.0
GAPFPR	4.4	8.7	7.5	107.4	938.9
GAPFQL	5.4	10.6	10.4	131.1	1292.9
ACDPAC	6.7	17.7	33.9	147.4	790.3
FMIN(1)	9.7	60.8	55.4	175.4	1468.7
FMIN(2)	18.0	239.0	181.0	1.4	1.4
FMIN(3)	14.1	23.7	22.2	514.0	4491.2
NLP	8.7	88.1	74.3	119.4	1045.3
SUMT	26.6	179.6	193.6	98.0	1087.0
DFP	8.4	42.0	46.3	76.3	827.7
FCDPAK	2.2	10.4	3.7	65.0	241.4

Table 29a: Efficiency related to accuracy.

Code	ET/A/B	NF/A/B	NG/A/B	NDF/A/B	NDG/A/B
OPRQP	2	3	3	8	9
XROP	1	2	2	7	6
VFO2AD	4	1	1	6	5
GRGA	5	9	9	11	8
OPT	12	21	22	1	10
GRG2(1)	9	13	12	9	11
GRG2(2)	16	20	21	1	1
VFO1A	6	5	6	17	12
LPNLP	11	12	10	18	21
SALQDR	7	10	7	24	22
SALQDF	13	22	19	3	2
SALMNF	22	4	13	25	25
CONMIN	16	19	20	10	13
BIAS(1)	14	17	17	13	14
BIAS(2)	23	24	24	4	3
FUNMIN	16	15	14	15	17
GAPFPR	8	5	5	19	18
GAPFQL	10	8	8	21	23
ACDPAC	15	11	15	22	15
FMIN(1)	21	18	18	23	24
FMIN(2)	25	26	25	5	4
FMIN(3)	24	14	11	26	26
NLP	20	23	23	20	19
SUMT	26	25	26	16	20
DFP	19	16	16	14	16
FCDPAK	3	7	4	12	7

Table 29b: Efficiency related to accuracy
(rank numbers).

Code	PNS	FFV	FVC	F
OPQRP	27.4	.48E-7	.15	1
XROP	37.5	.40E-6	.75	0
VF02AD	9.3	.43E-3	.65E-5	5
GRGA	42.0	.28E-3	.24E-10	3
OPT	65.3	.67E-2	.23E-7	7
GRG2(1)	21.2	.11E-4	.47E-4	0
GRG2(2)	29.1	.14E-6	.50E-4	9
VF01A	27.9	.21E-6	.42	13
LPNLP	36.7	.93E-8	.45E+1	10
SALQDR	76.6	.85E-4	.14E-4	13
SALQDF	76.9	.25E-3	.97E-5	15
SALMNF	23.1	.68E-6	.25E-3	18
CONMIN	71.7	.42E-4	.88E-4	14
BIAS(1)	35.1	.10E-5	.14E-5	4
BIAS(2)	53.7	.21E-5	.67E-5	13
FUNMIN	35.1	.99E-3	.58E-1	5
GAPFPR	46.4	.81E-3	.50E-9	24
GAPFQL	24.7	.12E-5	.19E-2	30
ACDPAC	14.4	.63E-5	.37E-4	8
FMIN(1)	52.9	.67E-7	.38E-2	0
FMIN(2)	89.6	.76E-6	.14E-3	0
FMIN(3)	31.3	.24E-5	.39E-1	15
NLP	18.4	.56E-7	.33	31
SUMT	79.2	.31E-3	.38E-6	19
DFP	54.3	.23	.81E-1	34
FCDPAK	51.7	.28E-7	.46E-2	2

Table 30a: Reliability.

Code	PNS	FFV	FVC	F
OPRQP	7	3	22	5
XROP	14	8	25	1
VF02AD	1	22	7	9
GRGA	15	19	1	7
OPT	21	25	3	11
GRG2(1)	4	16	11	1
GRG2(2)	9	6	12	13
VF01A	8	7	24	15
LPNLP	13	1	26	14
SALQDR	23	18	9	15
SALQDF	24	20	8	19
SALMNF	5	9	15	21
CONMIN	22	17	13	18
BIAS(1)	11	11	5	8
BIAS(2)	19	13	6	15
FUNMIN	11	24	20	9
GAPFPR	16	23	2	23
GAPFQL	6	12	16	24
ACDPAC	2	15	10	12
FMIN(1)	18	5	17	1
FMIN(2)	26	10	14	1
FMIN(3)	10	14	19	19
NLP	3	4	23	25
SUMT	25	21	4	22
DFP	20	26	21	26
FCDPAK	17	2	18	6

Table 30b: Reliability (rank numbers).

Code	PGS	GFV	GVC
OPRQP	20.9	-.33	.12E-7
XROP	20.7	-.24	.67E-8
VF02AD	26.5	-.90	.18E-8
GRGA	33.7	-.10E+2	.60E-10
OPT	51.3	-.32E+1	.78E-7
GRG2(1)	21.7	-.23E+1	.13E-4
GRG2(2)	26.5	-.24E+1	.40E-5
VF01A	14.5	-.59E-1	.14E-7
LPNLP	17.9	-.14E-2	.39E-5
SALQDR	66.0	-.12E+1	.42E-5
SALQDF	62.2	-.11E+1	.51E-5
SALMNF	26.6	-.44	.45E-6
CONMIN	7.1	-.13E-1	.22E-5
BIAS(1)	26.4	-.97E+1	.44E-5
BIAS(2)	20.4	-.59	.36E-5
FUNMIN	18.5	-.69E+1	.36E-5
GAPFPR	32.2	-.49	.24E-6
GAPFQL	5.3	-.57E-3	.23E-7
ACDPAC	23.2	-.25	.14E-5
FMIN(1)	15.0	-.19	.51E-5
FMIN(2)	40.0	-.32E-2	.24E-6
FMIN(3)	14.9	-.27E+1	.62E-5
NLP	22.5	-.38	.19E-5
SUMT	13.2	-.27E-1	.31E-7
DFP	31.7	-.83	.14E-5
FCDPAK	6.2	-.16E-1	.59E-8

Table 31a: Global convergence.

Code	PGS	GFV	GVC
OPRQP	15	16	5
XROP	16	18	4
VF02AD	9	10	2
GRGA	5	1	1
OPT	3	4	9
GRG2(1)	14	7	25
GRG2(2)	9	6	20
VF01A	22	20	6
LPNLP	19	25	19
SALQDR	1	8	21
SALQDF	2	9	23
SALMNF	8	14	12
CONMIN	24	23	16
BIAS(1)	11	2	22
BIAS(2)	17	12	17
FUNMIN	18	3	17
GAPFPR	6	13	10
GAPFQL	26	26	7
ACDPAC	12	17	13
FMIN(1)	20	19	23
FMIN(2)	4	24	10
FMIN(3)	21	5	26
NLP	13	15	15
SUMT	23	21	8
DFP	7	11	13
FCDPAK	25	22	3

Table 31b: Global convergence (rank numbers).

Code	E1	E2	R	G
OPRQP	4.14	6.68	9.03	14.08
XROP	3.34	5.00	12.96	15.12
VF02AD	4.18	4.24	5.71	8.44
GRGA	7.36	9.10	11.37	3.40
OPT	10.93	10.39	16.12	4.00
GRG2(1)	9.89	10.59	6.08	13.48
GRG2(2)	11.09	7.03	10.00	9.48
VF01A	9.33	11.65	12.25	19.52
LPNLP	14.40	16.78	14.34	20.68
SALQDR	13.61	18.27	18.27	5.36
SALQDF	9.85	7.81	19.46	6.48
SALMNF	21.31	20.53	10.38	10.16
CONMIN	15.61	13.98	18.97	22.76
BIAS(1)	14.22	14.44	9.29	9.80
BIAS(2)	15.34	10.00	15.30	15.60
FUNMIN	15.79	15.61	13.76	13.68
GAPFPR	11.85	14.49	15.44	8.44
GAPFQL	14.52	17.64	11.98	23.72
ACDPAC	16.12	16.86	6.85	13.52
FMIN(1)	21.58	21.92	13.15	20.08
FMIN(2)	16.87	11.19	17.21	10.32
FMIN(3)	23.19	22.37	13.86	17.00
NLP	20.22	20.44	11.09	13.80
SUMT	22.73	20.43	20.00	20.64
DFP	16.98	15.50	21.99	8.84
ACDPAK	5.95	8.07	13.45	21.52

Table 32: Weighted average rank numbers.

Class	1A	2A	3A	4A	5A	6A	7A	8A	9A	10A
FV	.86E-8	.61E-7	.16E-7	.25E-8	.24E-8	.28E-7	.17E-7	.20E-7	.12E-7	.62E-6
VC	.15E-7	.91E-7	.21E-7	.34E-9	.36E-8	.50E-8	.23E-7	.31E-7	.86E-9	.70E-6
KT	.26E-8	.18E-8	.91E-9	.23E-5	.28E-5	.25E-6	.13E-6	.13E-7	.90E-5	.12E-4
ED	9.37	10.43	9.66	6.70	6.61	7.30	7.19	9.08	4.90	6.15
ET	10.2	23.3	16.6	12.7	13.3	4.1	38.2	18.0	96.6	66.4
NF	90	77	28	37	33	50	30	26	131	172
NG	541	931	86	372	336	406	390	377	658	3099
NDF	53	50	28	30	29	36	28	26	71	99
NDG	321	604	84	300	296	293	364	377	360	1799
PNS	60.0	66.7	25.9	0	0	0	26.7	0	20.0	20.0
FFV	.22E-10	.72E-5	.61E-9	-	-	-	.11	-	.0	.20E-5
FVC	.47E+1	.42E+2	.20E-1	-	-	-	.53E-5	-	.43E+4	.15E-2
P	0	0	1	0	0	0	0	0	0	0
PGS	50.0	0	80.0	0	0	0	90.9	0	0	25.0
GFV	-.19	-	-.33E+1	-	-	-	-.18	-	-	-.15E-2
GVC	.44E-7	-	.20E-7	-	-	-	.18E-7	-	-	.49E-8

Table 33: Test results for OPRQP (general problems).

Class	1B	2B	3B	4B	5B	6B	7B	8B	9B	10B
FV	.17E-9	.11E-9	.21E-10	.10E-11	.65E-9	.58E-10	.10E-10	.47E-8	.15E-8	.39E-9
VC	.16E-9	.12E-9	.18E-10	.73E-11	.12E-8	.81E-10	.16E-11	.73E-8	.22E-8	.60E-8
KT	.15E-6	.19E-7	.57E-8	.87E-8	.87E-6	.74E-7	.16E-6	.28E-8	.22E-8	.11E-8
ED	7.77	8.54	9.02	8.85	7.29	8.08	6.26	9.67	9.66	9.38
A	8.54	9.03	9.67	10.01	7.86	8.89	8.96	8.67	8.95	8.74
ET	13.4	13.3	14.8	12.5	11.5	13.7	26.9	12.0	12.7	11.9
NF	31	32	35	36	26	32	82	25	27	25
NG	310	321	357	368	265	327	825	257	275	258
NDF	28	27	31	25	24	29	53	25	27	25
NDG	281	278	311	255	245	290	538	255	270	257
ET/A	1.6	1.5	1.5	1.3	1.5	1.6	3.0	1.4	1.4	1.4
NF/A	3.6	3.6	3.7	3.7	3.4	3.7	9.1	3.0	3.1	3.0
NG/A	36.1	35.7	36.7	37.4	33.5	36.9	91.3	29.6	30.7	29.5
NDF/A	3.3	3.1	3.2	2.6	3.1	3.2	6.0	3.0	3.0	2.9
NDG/A	32.8	30.9	32.1	25.6	31.1	32.7	59.9	29.4	30.1	29.4
R	0	0	0	0	0	0	0	0	1	0

Table 34: Test results for OPRQP (degenerate, ill-conditioned, and indefinite problems).

Class	1A	2A	3A	4A	5A	6A	7A	8A	9A	10A
PV	.14E-6	.10E-8	.68E-8	.87E-7	.18E-7	.70E-7	.40E-7	.13E-7	.28E-4	.13E-6
VC	.50E-7	.26E-7	.93E-8	.63E-8	.79E-8	.14E-7	.53E-7	.20E-7	.12E-6	.22E-6
KT	.39E-6	.23E-8	.48E-8	.13E-3	.48E-4	.34E-4	.63E-6	.91E-8	.30E-2	.21E-6
ED	7.96	10.89	9.35	5.21	5.48	5.56	6.83	9.20	2.81	8.03
ET	20.2	24.7	13.3	11.8	12.2	5.5	33.1	15.2	35.1	19.6
NF	192	87	20	25	24	64	23	19	29	30
NG	1155	1044	60	258	244	514	299	273	147	546
NDF	81	51	20	24	23	38	21	19	24	27
NDG	486	612	60	249	238	309	273	273	120	486
PNS	56.7	97.7	43.3	6.7	0	0	20.0	0	26.7	46.7
FFV	.10E-8	.13E-4	.36E-9	.14E-2	-	-	.78E-1	-	.13E-5	.54E-9
FVC	.13E+1	.28E+4	.16E+2	.25E-7	-	-	.31E-4	-	.54E+2	.39E-1
F	0	0	0	0	0	0	0	0	0	0
PGS	53.8	0	64.7	0	0	0	91.7	0	0	0
GFV	-.22	-	-.10E+1	-	-	-	-.19	-	-	-
GVC	.46E-8	-	.76E-8	-	-	-	.83E-8	-	-	-

Table 35: Test results for XROP (general problems).

Class	1B	2B	3B	4B	5B	6B	7B	8B	9B	10B
FV	.25E-7	.36E-7	.79E-7	.32E-6	.49E-8	.21E-6	.61E-6	.35E-7	.88E-7	.48E-7
VC	.14E-7	.27E-8	.31E-8	.24E-9	.85E-8	.30E-8	.75E-8	.79E-8	.53E-8	.79E-8
KT	.16E-3	.19E-3	.27E-3	.47E-3	.15E-4	.25E-3	.41E-3	.33E-4	.18E-3	.82E-4
ED	4.99	4.38	4.31	3.86	6.04	4.18	3.19	6.29	6.16	6.26
A	6.06	6.03	5.88	5.83	6.81	5.75	5.23	6.58	6.31	6.44
ET	7.7	8.2	7.8	6.7	7.4	8.5	8.7	8.1	8.2	8.3
NF	15	16	16	14	14	16	17	15	15	15
NG	155	165	160	142	141	165	171	156	157	156
NDF	14	15	15	12	14	16	16	15	15	15
NDG	146	156	150	127	141	162	166	155	156	155
ET/A	1.3	1.4	1.3	1.2	1.1	1.5	1.7	1.2	1.3	1.3
NF/A	2.6	2.7	2.7	2.5	2.1	2.8	3.4	2.4	2.5	2.4
NG/A	25.7	27.3	27.4	24.8	20.8	28.4	33.5	23.9	25.0	24.3
NDF/A	2.4	2.6	2.6	2.2	2.1	2.8	3.2	2.4	2.5	2.4
NDG/A	24.3	25.9	25.7	22.1	20.8	27.9	32.4	23.7	24.8	24.2
R	0	0	0	0	0	0	0	0	0	0

Table 36: Test results for XROP (degenerate, ill-conditioned, and indefinite problems).

Class	1A	2A	3A	4A	5A	6A	7A	8A	9A	10A
FV	.69E-7	.17E-11	.92E-8	.44E-7	.60E-8	.23E-7	.70E-8	.80E-8	–	.17E-8
VC	.26E-7	.18E-11	.21E-7	.11E-11	.10E-11	.76E-9	.12E-7	.18E-9	–	.13E-7
KT	.51E-5	.16E-11	.12E-4	.14E-3	.11E-3	.16E-3	.32E-4	.68E-5	–	.40E-5
ED	6.69	11.96	5.97	5.15	5.30	5.28	5.47	6.54	–	6.38
ET	4.7	5.3	11.2	23.3	31.0	23.2	55.0	53.7	–	153.5
NF	16	8	12	18	19	19	14	22	–	20
NG	96	96	37	184	191	157	192	320	–	362
NDF	16	8	12	18	19	19	14	22	–	20
NDG	96	96	37	184	191	157	192	320	–	362
PNS	6.7	10.0	23.3	3.3	0	0	0	0	0	8.3
FFV	.59E+1	.11E+1	.53E-2	.18E-2	–	–	–	–	–	.37E-2
FVC	.19E-3	.47E-6	.37E-1	.0	–	–	–	–	–	.37E-8
F	0	0	0	0	0	0	0	0	4	1
PGS	53.6	7.4	87.0	0	0	6.7	66.7	0	100.0	18.2
GFV	-.40E+1	-.20E+1	-.20E+1	–	–	-.17E-1	-.13	–	-.34E+2	-.16E-2
GVC	.91E-9	.37E-6	.14E-8	–	–	.56E-8	.57E-9	–	.31E-4	.75E-9

<u>Table 37</u>: Test results for VF02AD (general problems).

Class	1B	2B	3B	4B	5B	6B	7B	8B	9B	10B
FV	.25E-7	.68E-7	.93E-7	.12E-6	.47E-7	.78E-7	.42E-6	.15E-6	.16E-6	.16E-6
VC	.14E-11	.0	.0	.0	.0	.12E-11	.0	.0	.0	.0
KT	.20E-3	.26E-3	.32E-3	.30E-3	.25E-3	.17E-3	.25E-3	.17E-3	.16E-3	.19E-3
ED	4.77	4.27	4.23	3.96	4.82	4.35	3.19	6.48	6.45	6.45
A	6.98	6.76	6.69	6.60	6.94	6.78	6.29	7.27	7.26	7.25
ET	23.0	22.5	21.1	24.0	14.3	22.8	22.8	13.2	12.5	12.7
NF	13	13	12	14	8	12	13	10	9	10
NG	131	131	123	141	80	127	130	100	98	100
NDF	13	13	12	14	8	12	13	10	9	10
NDG	131	131	123	141	80	127	130	100	98	100
ET/A	3.3.	3.3	3.1	3.6	2.0	3.3	3.6	1.8	1.7	1.7
NF/A	1.9	2.0	1.9	2.1	1.1	1.9	2.1	1.3	1.3	1.4
NG/A	18.8	19.4	18.4	21.3	11.4	18.5	20.7	13.5	13.4	13.6
NDF/A	1.9	2.0	1.9	2.1	1.1	1.9	2.1	1.3	1.3	1.4
NDG/A	18.8	19.4	18.4	21.3	11.4	18.5	20.7	13.5	13.4	13.6
R	0	0	0	0	0	0	0	0	0	0

<u>Table 38</u>: Test results for VF02AD (degenerate, ill-conditioned, and indefinite problems).

Class	1A	2A	3A	4A	5A	6A	7A	8A	9A	10A
FV	.83E-7	.10E-7	.52E-6	.38E-4	.75E-4	.26E-4	.33E-10	.21E-3	.50E-4	.13E-4
VC	.25E-9	.97E-11	.25E-9	.11E-11	.10E-11	.15E-11	.21E-10	.14E-11	.24E-11	.59E-11
KT	.28E-3	.37E-8	.97E-3	.41E-2	.65E-2	.71E-2	.40E-9	.38E-1	.59E-2	.99E-2
ED	5.25	9.93	4.11	3.35	2.90	3.86	11.53	2.77	2.81	4.08
ET	16.6	51.0	25.5	52.2	33.4	6.3	7.2	41.7	89.2	47.7
NF	137	171	99	286	258	146	10	165	163	201
NG	1851	3985	457	4363	2077	1526	273	3236	1585	5670
NDF	61	72	34	78	76	50	2	67	79	70
NDG	168	367	57	456	470	232	39	432	260	641
PNS	3.3	24.7	3.3	3.3	3.3	20.0	24.7	0	0	40.0
FFV	.33E+1	.62	.65E-3	.17E-2	.80E-3	.11E+1	.13E-2	-	-	.61E-10
FVC	.27E-11	.0	.23E-9	.13E-11	.0	.17E-4	.14E-11	-	-	.26E+1
F	0	0	0	0	0	0	0	0	3	0
PGS	65.5	13.6	86.2	0	0	4.2	90.9	0	0	0
GFV	-.69E+1	-.20E+1	-.14E+3	-	-	-.17E-1	-.11	-	-	-
GVC	.21E-9	.0	.15E-9	-	-	.0	.31E-11	-	-	-

Table 39: Test results for GRGA (general problems).

Class	1B	2B	3B	4B	5B	6B	7B	8B	9B	10B
FV	.12E-3	.15E-3	.10E-3	.42E-4	.22E-3	.42E-3	.36E-3	.87E-6	.83E-6	.88E-6
VC	.11E-11	.14E-11	.0	.0	.0	.11E-11	.20E-11	.31E-10	.23E-10	.26E-10
KT	.13E-1	.19E-1	.15E-1	.73E-2	.17E-1	.23E-1	.18E-1	.30E-3	.31E-3	.34E-3
ED	2.97	2.83	2.98	2.88	3.04	2.64	2.38	5.74	5.76	5.76
A	5.18	5.05	5.20	5.35	5.11	4.91	4.82	6.46	6.50	6.47
ET	18.2	23.1	21.5	39.1	10.4	8.3	6.7	12.0	11.9	11.9
NF	109	137	137	184	57	42	32	62	62	62
NG	1135	1786	1326	3816	551	448	385	662	663	663
NDF	40	41	45	54	23	18	12	32	32	32
NDG	228	240	273	292	145	117	91	160	158	158
ET/A	3.5	4.7	4.1	7.6	2.0	1.7	1.4	1.9	1.8	1.8
NF/A	21.0	27.5	26.6	35.5	11.1	8.6	6.8	9.6	9.6	9.6
NG/A	220.6	364.0	255.5	748.1	107.3	91.4	80.8	102.2	101.7	102.2
NDF/A	7.7	8.4	8.6	10.5	4.5	3.7	2.7	5.0	4.9	4.9
NDG/A	44.3	47.7	52.6	55.7	28.2	23.8	19.0	24.7	24.3	24.4
R	0	0	0	0	0	1	0	0	0	0

Table 40: Test results for GRGA (degenerate, ill-conditioned, and indefinite problems).

Class	1A	2A	3A	4A	5A	6A	7A	8A	9A	10A
FV	.90E-7	.25E-7	.42E-4	.24E-3	.21E-3	.17E-4	-	.16E-4	-	.14E-4
VC	.15E-7	.55E-7	.55E-6	.33E-8	.11E-9	.39E-10	-	.62E-7	-	.20E-7
KT	.10E-3	.59E-8	.82E-2	.59E-2	.12E-1	.42E-2	-	.11E-1	-	.10E-1
ED	5.69	10.20	2.94	2.87	2.53	3.75	-	3.21	-	3.85
ET	16.5	52.4	60.1	58.4	56.9	19.0	-	99.3	-	168.2
NF	155	116	420	655	717	760	-	831	-	1372
NG	2325	5370	1483	5231	5531	5381	-	11104	-	21190
NDF	0	0	0	0	0	0	-	0	-	0
NDG	45	99	63	350	381	273	-	373	-	567
PNS	50.0	70.8	23.7	66.7	56.7	16.7	51.7	13.7	100.0	46.7
FFV	.50E-2	.51E+3	.30E-10	.29E-1	.11E-1	.18	.17E-2	.11E-1	.11E+1	.65
FVC	.14	.62E-9	.21	.88E-10	.13E-10	.21E-7	.68E-2	.25E-9	.87E-4	.22E-9
F	2	2	0	0	0	0	1	0	2	0
PGS	83.3	42.9	78.3	0	0	4.0	100.0	0	0	0
GFV	-.32E+2	-.20E+1	-.43E+1	-	-	-.17E-1	-.16	-	-	-
GVC	.12E-7	.16E-9	.67E-6	-	-	.58E-10	.18E-6	-	-	-

Table 41: Test results for OPT (general problems).

Class	1B	2B	3B	4B	5B	6B	7B	8B	9B	10B
FV	.28E-3	.40E-3	.25E-3	.26E-3	.53E-4	.38E-4	.13E-3	.29E-5	.52E-5	.29E-5
VC	.26E-6	.10E-8	.13E-8	.25E-9	.51E-9	.90E-9	.44E-8	.17E-7	.23E-8	.58E-8
KT	.18E-1	.26E-1	.21E-1	.22E-1	.77E-2	.62E-2	.11E-1	.62E-3	.98E-3	.63E-3
ED	2.77	2.66	2.68	2.71	3.12	3.22	2.63	5.40	5.26	5.38
A	3.66	4.16	4.21	4.38	4.70	4.72	4.21	5.48	5.55	5.59
ET	25.4	38.2	30.7	52.7	38.2	37.2	16.6	25.3	24.7	21.4
NF	310	484	373	675	483	478	182	292	288	246
NG	2388	3378	2936	4996	3865	3218	1732	2276	2218	1927
NDF	0	0	0	0	0	0	0	0	0	0
NDG	148	248	171	298	185	254	64	150	153	138
ET/A	7.1	8.1	7.4	11.9	7.9	8.0	3.7	4.6	4.3	3.7
NF/A	86.4	100.7	89.7	152.6	100.1	102.6	40.2	53.1	50.2	42.6
NG/A	676.9	728.7	725.8	1139.4	805.3	696.9	386.9	414.5	388.9	336.1
NDF/A	.0	.0	.0	.0	.0	.0	.0	.0	.0	.0
NDG/A	40.4	51.3	38.5	66.5	38.5	53.7	14.4	27.3	26.9	24.0
R	2	2	2	2	0	0	1	0	0	0

Table 42: Test results for OPT (degenerate, ill-conditioned, and indefinite problems).

Class	1A	2A	3A	4A	5A	6A	7A	8A	9A	10A
FV	.52E-4	.48E-6	.59E-6	.17E-7	.81E-8	.32E-5	.26E-4	.47E-4	.20E-6	.16E-4
VC	.29E-4	.12E-6	.33E-6	.75E-9	.92E-9	.44E-6	.12E-3	.34E-7	.13E-6	.49E-4
KT	.25E-2	.70E-7	.11E-3	.25E-4	.16E-4	.17E-3	.17E-2	.98E-3	.83E-4	.48E-3
ED	4.32	8.86	5.03	5.91	5.98	5.07	3.59	4.19	4.90	5.06
ET	12.0	24.8	47.9	33.7	33.8	6.5	146.3	45.7	216.5	172.7
NF	183	141	314	229	251	153	445	209	693	948
NG	1100	1703	2175	2295	2519	1226	5785	2938	3465	17076
NDF	15	13	37	28	32	27	50	37	109	104
NDG	93	165	245	286	329	218	650	519	550	1889
PNS	13.3	13.3	16.7	6.3	0	0	33.3	0	26.3	6.7
FFV	.28E-1	.17E+3	.45E+6	.17E-2	-	-	.34E-2	-	.83E-9	.37E-2
FVC	.19E-3	.10E-4	.36E-4	.12E-8	-	-	.14E-4	-	.22E-1	.22E-5
F	0	0	0	0	0	0	0	0	0	0
PGS	65.4	3.8	68.0	0	0	0	80.0	0	0	0
GFV	-.15E+1	-.20E+1	-.14E+2	-	-	-	-.23	-	-	-
GVC	.53E-5	.55E-4	.28E-4	-	-	-	.42E-4	-	-	-

Table 43: Test results for GRG2(1) (general problems).

Class	1B	2B	3B	4B	5B	6B	7B	8B	9B	10B
FV	.73E-8	.39E-8	.28E-7	.21E-8	.28E-8	.26E-6	.89E-6	.35E-5	.61E-5	.37E-5
VC	.32E-10	.54E-10	.11E-8	.41E-10	.96E-11	.20E-8	.37E-10	.98E-6	.11E-6	.11E-5
KT	.95E-5	.77E-4	.66E-4	.48E-4	.46E-5	.27E-3	.99E-4	.31E-4	.70E-4	.10E-4
ED	6.11	5.38	5.18	5.11	6.68	4.16	2.76	5.42	5.31	5.42
A	7.44	7.04	6.46	7.13	7.89	5.75	5.81	5.35	5.40	5.39
ET	17.3	17.1	15.2	19.4	16.6	18.2	22.2	8.5	8.5	8.5
NF	122	119	103	143	122	132	179	37	38	38
NG	1220	1198	1036	1438	1226	1323.	1793	378	380	382
NDF	15	15	14	16	14	15	16	10	10	10
NDG	157	157	146	163	141	157	160	106	106	106
ET/A	2.3	2.5	2.4	2.7	2.1	3.9	4.0	1.6	1.6	1.6
NF/A	16.1	17.3	16.2	19.7	15.2	29.0	32.0	7.1	7.3	7.1
NG/A	160.7	172.5	162.0	197.2	152.3	289.9	319.9	71.1	73.2	71.0
NDF/A	2.1	2.3	2.3	2.3	1.8	3.2	2.9	2.0	2 0	2.0
NDG/A	21.0	22.7	23.0	22.7	17.9	31.7	28.8	19.9	20.4	19.7
R	0	0	0	0	0	0	0	0	0	0

Table 44: Test results for GRG2(1) (degenerate, ill-conditioned, and indefinite problems).

Class	1A	2A	3A	4A	5A	6A	7A	8A	9A	10A
FV	.12E-3	.29E-4	.27E-4	.69E-5	.68E-7	.15E-5	.82E-6	.25E-6	.17E-4	.91E-5
VC	.13E-3	.68E-4	.52E-4	.28E-9	.87E-10	.10E-6	.16E-5	.12E-7	.16E-6	.24E-4
KT	.70E-3	.68E-5	.40E-3	.31E-2	.40E-3	.83E-3	.13E-3	.23E-3	.40E-2	.12E-2
ED	4.87	7.23	4.23	3.97	4.71	4.59	4.70	4.87	3.13	4.97
ET	13.8	18.7	43.8	49.6	54.9	13.0	238.6	134.6	216.9	350.7
NF	245	128	354	471	670	466	1300	1104	1888	2565
NG	1473	1542	1063	4715	6707	3734	16900	15461	9444	46183
NDF	0	0	0	0	0	0	0	0	0	0
NDG	0	0	0	0	0	0	0	0	0	0
PNS	56.7	60.0	13.3	6.7	0	0	8.3	0	26.7	33.3
FPV	.76E-10	.14E-3	.33E+3	.17E-2	-	-	.20E-1	-	.84E-9	.30E-2
FVC	.38E-2	.84E-2	.25E-4	.12E-8	-	-	.10E-5	-	.44E-2	.80E-5
F	0	5	0	0	0	0	1	0	0	3
PGS	84.6	0	88.5	0	0	0	90.9	0	0	0
GFV	-.40	-	-.26E+2	-	-	-	-.12	-	-	-
GVC	.11E-4	-	.14E-4	-	-	-	.67E-6	-	-	-

Table 45: Test results for GRG2(2) (general problems).

Class	1B	2B	3B	4B	5B	6B	7B	8B	9B	10B
FV	.61E-7	.29E-6	.30E-6	.39E-6	.34E-7	.18E-6	.19E-5	.34E-5	.59E-5	.37E-5
VC	.13E-9	.20E-9	.74E-9	.21E-10	.25E-10	.31E-9	.47E-9	.99E-6	.11E-6	.11E-5
KT	.54E-3	.88E-3	.90E-3	.11E-2	.20E-3	.28E-3	.35E-3	.36E-3	52E-3	.34E-3
ED	4.66	4.24	4.16	4.08	4.92	4.26	2.69	.5.15	5.12	5.11
A	6.26	5.88	5.71	6.04	6.67	6.02	5.30	5.02	5.15	4.99
ET	28.2	23.2	21.6	26.6	23.8	33.6	19.4	13.5	14.0	13.7
NF	333	273	255	315	283	406	231	154	159	155
NG	3335	2737	2550	3158	2831	4061	2313	1547	1596	1551
NDF	0	0	0	0	0	0	0	0	0	0
NDG	0	0	0	0	0	0	0	0	0	0
ET/A	4.5	4.1	3.9	4.4	3.6	5.5	4.2	2.7	2.8	2.8
NF/A	52.7	47.9	45.7	52.5	42.4	66.8	50.3	31.1	31.7	31.1
NG/A	526.8	478.6	456.8	525.3	423.9	667.7	503.0	311.1	317.1	311.4
NDF/A	0.0	0.0	0.0	0.0	0.0	0.0	0.0	0.0	0.0	0.0
NDG/A	0.0	0.0	0.0	0.0	0.0	0.0	0.0	0.0	0.0	0.0
R	0	0	0	0	0	0	0	0	0	0

Table 46: Test results for GRG2(2) (degenerate, ill-conditioned, and indefinite problems).

Class	1A	2A	3A	4A	5A	6A	7A	8A	9A	10A
FV	.23E-7	.38E-5	.13E-8	.32E-9	.21E-8	.13E-7	.23E-8	.14E-7	.11E-7	.82E-8
VC	.70E-8	.64E-5	.22E-8	.13E-9	.85E-10	.55E-9	.29E-8	.40E-7	.39E-7	.38E-7
KT	.11E-6	.39E-7	.62E-8	.70E-6	.15E-5	.17E-5	.56E-7	.57E-6	.64E-6	.85E-6
ED	7.93	9.00	9.21	7.29	6.86	6.68	8.03	7.58	6.62	7.61
ET	33.6	91.4	46.0	17.8	33.2	5.2	121.1	52.5	143.0	57.4
NF	263	403	80	107	184	98	162	150	179	204
NG	1580	4842	242	1073	1848	785	2109	2103	899	3685
NDF	263	403	80	107	184	98	162	150	179	204
NDG	850	1979	242	226	481	252	974	942	475	2007
PNS	33.3	85.2	50.0	3.3	0	0	33.3	0	26.3	11.1
FFV	.23E-3	.88E-7	.28E-1	.17E-2	–	–	.13E-7	–	.0	0.
PVC	.32E+1	.12E+3	.15E+1	.0	–	–	.23E-4	–	.19E+5	.32E-1
F	·2	1	6	0	0	2	0	0	0	2
PGS	43.8	0	33.3	0	0	4.2	60.0	0	0	0
GFV	-.94	–	-.29	–	–	-.17E-1	-.37	–	–	–
GVC	.72E-7	–	.27E-8	–	–	.21E-9	.43E-8	–	–	–

Table 47: Test results for VF01A (general problems).

Class	1B	2B	3B	4B	5B	6B	7B	8B	9B	10B
FV	.11E-8	.17E-9	.10E-8	.11E-11	.28E-9	.27E-9	.44E-4	.19E-8	.34E-8	.28E-9
VC	.11E-8	.55E-10	.52E-9	.23E-9	.22E-9	.42E-9	.35E-4	.34E-8	.48E-8	.42E-10
KT	.25E-6	.48E-6	.11E-5	.20E-7	.48E-6	.22E-6	.16E-3	.54E-7	.10E-7	.40E-7
ED	7.66	7.12	6.80	7.57	7.58	7.42	2.75	9.20	9.01	8.91
A	8.05	8.37	7.75	9.21	8.28	8.26	3.84	8.42	8.45	9.06
ET	31.7	29.5	38.9	7.4	20.7	39.9	62.3	37.4	27.2	27.0
NF	144	151	208	46	103	207	330	165	119	117
NG	1447	1511	2087	467	1033	2072	3303	1651	1196	1170
NDF	144	151	208	46	103	207	330	165	119	117
NDG	482	376	512	72	306	584	907	633	453	436
ET/A	3.9	3.6	5.2	0.8	2.5	4.9	16.7	4.9	3.8	3.0
NF/A	18.1	18.2	27.9	5.2	12.6	25.2	90.1	21.6	16.9	13.1
NG/A	181.4	181.8	278.6	51.7	125.7	252.0	901.4	215.8	168.5	130.7
NDF/A	18.1	18.2	27.9	5.2	12.6	25.2	90.1	21.6	16.9	13.1
NDG/A	60.0	44.8	67.6	8.1	37.3	71.0	239.9	83.2	64.3	48.8
R	0	1	1	0	0	0	0	0	0	0

Table 48: Test results for VF01A (degenerate, ill-conditioned, and indefinite problems).

Class	1A	2A	3A	4A	5A	6A	7A	8A	9A	10A
FV	.99E-7	.41E-6	.50E-8	.11E-4	.15E-4	.15E-6	-	.90E-7	.54E-6	.51E-6
VC	.25E-6	.78E-6	.17E-7	.34E-4	.38E-4	.12E-7	-	.32E-6	.15E-5	.25E-5
KT	.83E-6	.76E-6	.96E-7	.91E-3	.86E-3	.26E-5	-	.38E-5	.51E-4	.55E-4
ED	6.91	8.25	8.21	3.30	3.33	6.15	-	6.76	3.44	5.68
ET	36.5	129.3	19.5	54.6	53.9	7.7	-	63.4	171.7	72.4
NF	469	644	57	226	194	168	-	225	364	352
NG	2813	7731	172	2261	1948	1347	-	3155	1822	6352
NDF	154	147	28	113	113	62	-	83	121	117
NDG	929	1764	85	1132	1138	502	-	1170	608	2123
PNS	29.2	87.8	72.2	3.3	0	0	93.3	0	20.0	26.7
FFV	.12E-3	.24E-2	.0	.34E-2	-	-	.29E-9	-	.0	.26E-2
FVC	.82	.67E+2	.20E+12	.0	-	-	.54E-1	-	.26E+2	.86E-5
F	2	4	4	0	0	0	0	0	0	0
PGS	41.2	0	0	10.3	0	10.0	100.0	0	0	0
GFV	-.21	-	-	-.15E-2	-	-.17E-1	-.87E-1	-	-	-
GVC	.12E-5	-	-	.22E-3	-	.60E-7	.15E-5	-	-	-

Table 49: Test results for LPNLP (general problems).

Class	1B	2B	3B	4B	5B	6B	7B	8B	9B	10B
FV	.21E-4	.21E-5	.14E-4	.43E-11	.61E-5	.14E-4	.19E-3	.86E-8	.17E-8	.62E-8
VC	.40E-4	.35E-5	.36E-4	.99E-8	.19E-4	.35E-4	.22E-3	.44E-8	.11E-7	.14E-7
KT	.32E-2	.27E-3	.23E-2	.13E-6	.19E-2	.12E-2	.16E-2	.18E-7	.17E-7	.39E-7
ED	3.09	3.60	2.82	5.76	3.87	3.32	2.09	8.02	8.12	7.66
A	3.67	4.58	3.69	8.00	4.13	3.89	3.07	8.05	8.15	7.79
ET	97.3	66.9	95.9	9.6	76.2	86.2	116.5	24.4	26.2	24.6
NF	386	272	410	41	255	425	631	119	137	124
NG	3863	2722	4103	412	2555	4250	6313	1197	1372	1240
NDF	187	127	179	17	160	157	187	42	43	42
NDG	1870	1278	1793	173	1607	1571	1878	423	435	421
ET/A	26.9	16.5	26.9	1.2	19.3	23.1	37.9	3.1	3.3	3.1
NF/A	107.2	67.3	116.4	5.1	64.3	113.8	206.1	15.3	17.1	15.8
NG/A	1072.0	672.8	1164.2	50.8	643.2	1137.6	2060.6	152.9	171.2	158.0
NDF/A	51.8	31.4	49.9	2.1	40.9	42.3	61.1	5.3	5.4	5.4
NDG/A	518.0	313.6	499.8	21.5	408.5	422.8	611.4	53.4	53.9	53.8
R	0	0	0	0	0	0	0	0	0	0

Table 50: Test results for LPNLP (degenerate, ill-conditioned, and indefinite problems).

Class	1A	2A	3A	4A	5A	6A	7A	8A	9A	10A
FV	.49E-4	-	.34E-6	.85E-4	.13E-3	.12E-3	-	.18E-3	.12E-3	.82E-4
VC	.61E-5	-	.54E-6	.45E-5	.12E-7	.31E-5	-	.11E-4	.26E-4	.55E-5
KT	.49E-2	-	.81E-4	.90E-2	.70E-2	.17E-1	-	.35E-1	.56E-2	.24E-1
ED	3.63	-	5.40	2.81	2.64	3.01	-	2.87	2.30	3.27
ET	48.0	-	27.1	23.3	27.3	5.4	-	69.2	138.6	54.3
NF	353	-	47	59	65	55	-	117	115	135
NG	2118	-	143	591	652	445	-	1640	578	2434
NDF	350	-	46	55	61	54	-	115	112	132
NDG	2105	-	140	554	618	437	-	1621	562	2376
PNS	16.7	-	36.7	40.0	53.3	85.2	86.7	66.7	55.6	46.7
FFV	.11E-3	-	.83E-9	.21E-1	.24E-2	.10E-2	.94E-5	.46E-2	.45E-2	.41E-2
FVC	.20	-	.29E-1	.13E-5	.26E-5	.17E-5	.36E-2	.14E-4	.43E-6	.81E-4
F	0	10	0	0	0	1	0	0	2	0
PGS	60.0	-	73.7	5.6	0	25.0	100.0	0	0	0
GFV	-.11E+1	-	-.39E+1	-.74E-4	-	-.17E-1	-.18	-	-	-
GVC	.76E-5	-	.32E-5	.47E-3	-	.71E-6	.17E-4	-	-	-

Table 51: Test results for SALQDR (general problems).

Class	1B	2B	3B	4B	5B	6B	7B	8B	9B	10B
FV	.27E-3	.15E-3	.61E-4	.13E-4	.96E-4	.33E-3	.58E-3	.52E-3	.10E-3	.12E-3
VC	.10E-5	.14E-4	.21E-4	.20E-4	.15E-5	.14E-4	.77E-4	.34E-8	.83E-7	.38E-7
KT	.26E-1	.12E-1	.58E-2	.19E-2	.13E-1	.23E-1	.23E-1	.98E-2	.41E-2	.36E-2
ED	2.38	2.50	2.43	2.62	3.09	2.40	2.13	4.14	4.69	4.60
A	3.38	3.28	3.39	3.73	3.71	3.09	2.80	4.48	4.54	4.59
ET	23.6	36.5	28.7	25.5	21.0	26.3	28.7	19.0	21.3	21.2
NF	57	85	69	59	48	62	68	41	46	47
NG	570	856	693	598	487	628	683	415	461	470
NDF	50	79	63	55	44	56	60	40	45	45
NDG	500	798	630	553	447	566	606	405	451	455
ET/A	7.4	11.1	8.7	7.3	6.0	8.7	10.2	4.3	4.8	4.6
NF/A	17.6	26.1	21.0	17.2	13.9	21.0	24.4	9.4	10.3	10.3
NG/A	175.8	261.3	209.5	171.9	138.7	210.0	244.2	94.2	103.0	102.8
NDF/A	15.8	24.3	19.1	15.9	12.8	18.8	21.6	9.2	10.1	9.9
NDG/A	155.7	243.3	190.5	158.9	127.8	188.3	216.1	92.0	100.8	99.2
R	2	1	1	0	0	0	0	1	1	0

Table 52: Test results for SALQDR (degenerate, ill-conditioned, and indefinite problems).

Class	1A	2A	3A	4A	5A	6A	7A	8A	9A	10A
FV	.53E-4	-	.91E-7	.71E-4	.23E-3	.17E-3	-	.19E-3	.17E-3	.22E-4
VC	.45E-5	-	.38E-6	.40E-5	.81E-8	.29E-6	-	.15E-4	.75E-5	.13E-5
KT	.89E-2	-	.41E-4	.61E-2	.12E-1	.12E-1	-	.37E-1	.11E-1	.12E-1
ED	3.39	-	5.80	2.91	2.59	3.10	-	2.71	2.37	3.54
ET	128.2	-	49.5	43.2	43.4	20.4	-	191.1	160.9	388.2
NF	2798	-	493	509	639	866	-	1910	1658	3265
NG	16645	-	1303	4430	5688	6253	-	25358	7605	55295
NDF	0	-	0	0	0	0	-	0	0	0
NDG	0	-	0	0	0	0	-	0	0	0
PNS	20.0	-	33.3	53.3	46.7	59.3	93.3	60.0	44.4	66.7
FFV	.17E-4	-	.14E-8	.89E-2	.33E-2	.60E-2	.37E-2	.36E-2	.42E-2	.37E-2
FVC	.73	-	.24E-1	.99E-7	.54E-5	.52E-5	.11E-3	.31E-4	.29E-6	.28E-3
F	0	10	0	0	0	1	0	0	2	2
PGS	62.5	-	75.0	7.1	0	0	100.0	0	0	0
GFV	-.11E+1	-	-.46E+1	-.14E-2	-	-	-.64	-	-	-
GVC	.53E-5	-	.13E-4	.19E-3	-	-	.30E-6	-	-	-

Table 53: Test results for SALQDF (general problems).

Class	1B	2B	3B	4B	5B	6B	7B	8B	9B	10B
FV	.24E-3	.17E-3	.38E-4	.67E-5	.21E-3	.11E-2	.88E-3	.48E-5	.89E-5	.56E-5
VC	.44E-5	.87E-5	.87E-6	.23E-5	.80E-5	.23E-6	.96E-4	.44E-7	.72E-6	.23E-8
KT	.17E-1	.13E-1	.68E-2	.23E-2	.14E-1	.41E-1	.38E-1	.22E-3	.86E-3	.30E-3
ED	2.45	2.46	2.68	3.03	2.91	2.38	2.13	5.57	5.17	5.46
A	3.30	3.28	3.83	4.12	3.39	3.34	2.66	5.48	4.86	5.72
ET	44.4	42.9	37.8	43.1	35.3	38.3	41.5	36.6	37.5	41.3
NF	646	617	563	633	538	573	610	544	533	582
NG	5765	5445	4908	5575	4642	5051	5452	4744	4642	5142
NDF	0	0	0	0	0	0	0	0	0	0
NDG	0	0	0	0	0	0	0	0	0	0
ET/A	14.0	13.4	10.2	10.9	10.7	12.5	16.2	7.0	8.0	8.0
NF/A	203.4	192.3	151.7	161.2	163.6	185.6	237.3	104.2	115.4	113.0
NG/A	1814.3	1696.9	1321.2	1419.6	1412.9	1640.6	2127.5	899.8	994.2	997.3
NDF/A	.0	.0	.0	.0	.0	.0	.0	.0	.0	.0
NDG/A	.0	.0	.0	.0	.0	.0	.0	.0	.0	.0
R	0	1	1	0	0	0	0	1	2	2

Table 54: Test results for SALQDF (degenerate, ill-conditioned, and indefinite problems).

Class	1A	2A	3A	4A	5A	6A	7A	8A	9A	10A
FV	.52E-5	-	.26E-5	.13E-5	.14E-5	.37E-9	-	.43E-9	.24E-5	.16E-8
VC	.15E-4	-	.21E-4	.86E-6	.20E-5	.39E-10	-	.21E-9	.19E-5	.53E-8
KT	.51E-3	-	.17E-5	.31E-3	.47E-3	.48E-7	-	.28E-6	.76E-3	.37E-6
ED	4.61	-	6.62	4.06	3.96	8.26	-	7.81	2.70	7.78
ET	147.6	-	66.4	79.5	110.5	22.3	-	285.0	409.9	288.7
NF	429	-	24	35	46	61	-	74	28	70
NG	7235	-	356	2054	2623	2579	-	7347	1686	16690
NDF	1205	-	118	205	262	322	-	524	337	927
NDG	7235	-	356	2054	2623	2579	-	7347	1686	16690
PNS	20.0	-	20.0	10.0	0	11.1	88.9	0	0	20.0
PPV	.26E-5	-	.0	.17E-2	-	.18E-2	.21E-3	-	-	.43E-2
FVC	.14E-1	-	.14E+1	.14E-11	-	.30E-4	.20E-2	-	-	.17E-3
F	0	10	0	0	0	1	2	1	4	0
PGS	62.5	-	79.2	0	0	12.5	100.0	0	0	0
GFV	-.11E+1	-	-.29E+1	-	-	-.17E-1	-.87E-1	-	-	-
GVC	.50E-6	-	.23E-4	-	-	.62E-10	.47E-6	-	-	-

Table 55: Test results for SALMNF (general problems).

Class	1B	2B	3B	4B	5B	6B	7B	8B	9B	10B
FV	.34E-4	.29E-5	.60E-5	.61E-9	.31E-5	.24E-5	.16E-3	.22E-11	.62E-11	.34E-11
VC	.43E-7	.66E-7	.50E-7	.59E-8	.50E-6	.23E-6	.78E-6	.13E-11	.13E-11	.13E-11
KT	.38E-2	.86E-3	.96E-3	.26E-5	.11E-2	.92E-3	.94E-2	.16E-10	.71E-10	.28E-10
ED	3.16	3.69	3.66	5.65	4.00	3.58	2.04	10.92	10.53	10.88
A	4.35	4.87	4.80	7.17	4.69	4.72	3.49	11.31	10.94	11.20
ET	122.8	109.0	110.2	39.7	92.5	104.1	80.8	56.7	55.7	55.0
NF	58	45	49	14	39	40	48	19	19	19
NG	2805	2432	2511	897	2157	2443	1867	1318	1300	1286
NDF	280	243	251	89	216	244	186	131	130	128
NDG	2805	2432	2511	897	2157	2443	1867	1318	1300	1286
ET/A	28.0	23.3	24.4	6.5	20.4	23.4	23.9	5.0	5.2	4.9
NF/A	12.9	9.9	11.0	2.4	8.9	8.9	14.7	1.7	1.8	1.7
NG/A	638.0	520.3	553.9	146.6	476.2	547.9	551.3	117.0	120.6	115.7
NDF/A	63.8	52.0	55.4	14.7	47.6	54.8	55.1	11.7	12.1	11.6
NDG/A	638.0	520.3	553.9	146.6	476.2	547.9	551.3	117.0	120.6	115.7
R	0	0	0	0	0	0	0	0	0	0

Table 56: Test results for SALMNF (degenerate, ill-conditioned, and indefinite problems).

Class	1A	2A	3A	4A	5A	6A	7A	8A	9A	10A
FV	.16E-3	-	.32E-7	.46E-6	.23E-6	.19E-5	-	.21E-5	.49E-5	-
VC	.12E-4	-	.12E-6	.63E-9	.18E-8	.56E-7	-	.22E-5	.13E-6	-
KT	.27E-1	-	.10E-6	.30E-3	.20E-3	.45E-3	-	.82E-3	.13E-2	-
ED	2.87	-	8.40	4.78	4.81	4.37	-	4.42	3.96	-
ET	76.5	-	56.4	94.4	97.2	26.7	-	196.8	165.8	-
NF	1462	-	368	799	953	921	-	1363	957	-
NG	8775	-	1104	7991	9530	7369	-	19094	4786	-
NDF	106	-	31	54	64	58	-	82	58	-
NDG	639	-	93	543	652	467	-	1160	292	-
PNS	83.3	100.0	76.7	40.0	26.7	37.5	-	13.3	65.6	-
FFV	.45E-5	.65E-4	.91E-7	.42E-1	.48E-2	.45E-1	-	.12E-2	.25E-3	-
FVC	.66E-2	.33E-1	.30	.25E-8	.12E-8	.14E-6	-	.11E-3	.19E-5	-
F	0	0	0	0	0	2	5	0	2	5
PGS	60.0	0	28.6	0	0	0	-	0	0	-
GFV	-.73E+1	-	-.82E-2	-	-	-	-	-	-	-
GVC	.19E-3	-	.11E-3	-	-	-	-	-	-	-

Table 57: Test results for CONMIN (general problems).

Class	1B	2B	3B	4B	5B	6B	7B	8B	9B	10B
FV	.89E-7	.25E-7	.31E-7	.31E-7	.63E-7	.18E-7	.17E-6	.81E-8	.44E-8	.37E-8
VC	.84E-8	.61E-8	.85E-8	.16E-10	.17E-6	.93E-8	.55E-8	.23E-9	.17E-9	.43E-8
KT	.13E-3	.12E-3	.15E-3	.16E-3	.97E-4	.21E-4	.34E-4	.13E-5	.20E-6	.17E-6
ED	4.88	4.35	4.62	4.28	5.27	5.59	3.23	7.58	7.80	7.72
A	5.97	6.02	6.00	6.59	5.81	6.52	5.68	7.55	8.16	7.82
ET	58.8	60.8	63.8	66.7	49.9	61.6	117.4	29.3	31.4	30.6
NF	458	505	535	574	401	492	1056	217	233	227
NG	4587	5057	5355	5743	4012	4923	10560	2176	2332	2275
NDF	40	43	45	48	38	42	63	18	20	19
NDG	405	431	455	480	381	420	638	187	202	196
ET/A	9.9	10.0	10.6	10.3	8.6	9.5	37.2	3.9	4.0	4.1
NF/A	77.5	83.5	90.2	89.1	69.1	76.2	347.0	29.3	29.8	30.9
NG/A	774.5	834.8	902.3	890.9	691.4	761.6	3460.1	292.9	297.9	308.6
NDF/A	6.8	7.1	7.6	7.4	6.6	6.5	18.3	2.5	2.5	2.6
NDG/A	68.3	71.3	76.4	74.3	65.8	65.3	183.4	24.6	25.3	26.1
R	0	0	0	0	0	0	0	0	0	0

Table 58: Test results for CONMIN (degenerate, ill-conditioned, and indefinite problems).

Class	1A	2A	3A	4A	5A	6A	7A	8A	9A	10A
FV	.27E-4	.14E-3	.21E-5	.91E-6	.19E-5	.90E-5	.55E-6	.49E-6	.76E-6	.14E-4
VC	.40E-5	.31E-3	.52E-5	.11E-5	.39E-5	.57E-6	.64E-5	.17E-5	.12E-5	.43E-6
KT	.13E-4	.73E-6	.14E-5	.20E-3	.22E-3	.66E-4	.41E-3	.36E-4	.40E-4	.12E-3
ED	5.70	7.39	6.86	4.31	4.18	4.58	4.10	5.79	4.98	5.18
ET	26.1	82.8	53.3	62.8	45.7	17.7	220.8	102.1	123.8	173.5
NF	416	604	317	458	372	617	691	601	543	1070
NG	2779	7788	1062	5228	4238	5533	10157	9515	3049	21733
NDF	45	43	35	59	48	71	87	75	65	133
NDG	270	516	107	593	485	575	1131	1062	328	2410
PNS	33.3	96.7	3.7	3.3	0	6.7	25.0	0	75.0	20.0
FFV	.59E-6	.21E-1	.0	.17E-2	-	.0	.0	-	.80E-5	.0
FVC	.17E-1	.30E-9	.10E-2	.0	-	.11E-2	.53E-2	-	.29E-3	.62E-2
F	1	0	1	0	0	0	1	0	1	0
PGS	66.7	0	76.9	0	0	10.7	66.7	0	0	16.7
GFV	-.14E+1	-	-.65E+3	-	-	-.17E-1	-.32	-	-	-.15E-2
GVC	.16E-4	-	.21E-4	-	-	.39E-10	.10E-3	-	-	.15E-4

Table 59: Test results for BIAS(1) (general problems).

Class	1B	2B	3B	4B	5B	6B	7B	8B	9B	10B
FV	.43E-4	.35E-5	.11E-4	.13E-10	.56E-5	.67E-5	.76E-5	.18E-4	.28E-5	.19E-6
VC	.58E-4	.93E-5	.14E-4	.35E-7	.12E-4	.18E-4	.16E-4	.23E-5	.57E-8	.12E-6
KT	.34E-2	.30E-3	.11E-2	.48E-6	.13E-2	.78E-3	.35E-4	.35E-4	.37E-5	.54E-6
ED	3.00	3.49	3.26	5.42	4.06	3.60	2.71	6.21	6.42	6.85
A	3.52	4.38	4.01	7.53	4.29	4.16	4.27	5.26	6.41	6.69
ET	73.2	83.1	81.0	16.6	75.6	66.2	106.1	29.2	28.0	24.4
NF	525	612	597	139	542	508	865	249	240	207
NG	6146	7107	6982	1576	6441	5856	9777	2794	2681	2301
NDF	83	92	94	16	91	72	108	27	25	20
NDG	830	920	945	165	912	722	1083	272	257	207
ET/A	21.8	20.5	21.0	2.2	18.1	16.8	26.3	5.7	4.6	3.6
NF/A	155.9	150.1	154.8	18.2	129.7	128.7	213.9	48.9	39.8	30.6
NG/A	1823.8	1742.6	1808.3	207.0	1539.2	1483.9	2419.6	547.7	443.0	339.6
NDF/A	24.7	22.8	24.4	2.2	21.9	18.4	26.9	5.4	4.2	3.1
NDG/A	246.6	227.6	244.4	22.0	218.8	183.7	268.9	53.9	42.3	30.8
R	0	0	0	0	0	0	0	1	1	1

Table 60: Test results for BIAS(1) (degenerate, ill-conditioned, and indefinite problems).

Class	1A	2A	3A	4A	5A	6A	7A	8A	9A	10A
FV	.49E-5	.15E-3	.73E-5	.54E-5	.26E-5	.25E-4	.17E-4	.33E-4	.40E-3	.59E-4
VC	.17E-5	.39E-3	.13E-4	.45E-6	.50E-5	.48E-5	.52E-4	.70E-5	.18E-4	.48E-4
KT	.81E-4	.17E-5	.67E-4	.10E-2	.33E-3	.22E-2	.43E-2	.14E-1	.18E-1	.13E-1
ED	5.15	7.16	5.19	3.63	4.03	3.17	3.06	3.15	2.44	3.34
ET	39.0	77.9	70.0	108.7	74.3	46.5	376.4	247.8	334.3	514.8
NF	837	681	658	1269	1082	2216	2417	2500	3471	4457
NG	5036	8196	1978	12730	10851	17756	31460	35062	17368	80308
NDF	0	0	0	0	0	0	0	0	0	0
NDG	0	0	0	0	0	0	0	0	0	0
PNS	22.2	95.3	8.3	10.0	3.3	33.3	22.2	0	66.7	33.3
FFV	.19E-2	.22E-1	.56E+2	.14E-5	.0	.81E-11	.64E-7	-	.64E-3	.37E-8
FVC	.22E-1	.20E-8	.68E-5	.70E-4	.12E-2	.42E-2	.16E-2	-	.49E-3	.82E-2
F	4	0	6	0	0	0	2	0	1	0
PGS	57.1	50.0	54.5	0	0	15.0	57.1	0	25.0	10.0
GFV	-.56	-.22E-1	-.85E+2	-	-	-.17E-1	-.24	-	-.20E+2	-.11E-2
GVC	.27E-4	.20E-8	.59E-5	-	-	.30E-5	.77E-4	-	.15E-3	.23E-3

Table 61: Test results for BIAS(2) (general problems).

Class	1B	2B	3B	4B	5B	6B	7B	8B	9B	10B
FV	.37E-4	.29E-4	.34E-4	.17E-6	.74E-6	.22E-4	.29E-4	.11E-4	.65E-5	.33E-5
VC	.87E-5	.36E-4	.55E-4	.38E-7	.54E-5	.15E-4	.25E-4	.27E-5	.11E-8	.51E-6
KT	.32E-2	.20E-2	.27E-2	.15E-3	.79E-3	.18E-2	.70E-3	.15E-3	.33E-3	.16E-3
ED	2.86	2.86	2.78	3.46	4.04	3.33	2.49	4.64	4.69	4.94
A	3.71	3.63	3.52	5.37	4.63	3.89	3.70	4.67	5.58	5.13
ET	138.1	164.7	159.9	96.5	131.8	140.6	173.7	67.0	66.6	64.1
NF	1737	2084	2024	1224	1640	1751	2187	838	826	795
NG	17416	20886	20275	12271	16452	17544	21904	8408	8287	7978
NDF	0	0	0	0	0	0	0	0	0	0
NDG	0	0	0	0	0	0	0	0	0	0
ET/A	39.1	47.7	47.3	20.3	29.3	36.9	49.2	14.4	12.8	12.7
NF/A	491.6	603.8	599.0	257.1	365.3	456.0	620.5	179.9	159.2	158.2
NG/A	4927.6	6049.0	5999.3	2576.9	3663.5	4606.9	6213.3	1803.7	1595.6	1585.7
NDF/A	0.0	0.0	0.0	0.0	0.0	0.0	0.0	0.0	0.0	0.0
NDG/A	0.0	0.0	0.0	0.0	0.0	0.0	0.0	0.0	0.0	0.0
R	0	0	0	0	0	1	2	1	0	1

Table 62: Test results for BIAS(2) (degenerate, ill-conditioned, and indefinite problems).

Class	1A	2A	3A	4A	5A	6A	7A	8A	9A	10A
FV	.22E-3	-	.18E-8	.31E-9	.15E-9	.15E-8	.51E-9	.29E-9	.20E-8	.16E-7
VC	.12E-4	-	.26E-8	.16E-9	.43E-9	.13E-9	.25E-8	.12E-8	.18E-8	.20E-8
KT	.24E-1	-	.13E-5	.80E-5	.47E-5	.97E-5	.36E-5	.12E-4	.71E-5	.95E-4
ED	3.41	-	6.82	6.35	6.48	6.22	6.47	6.30	5.10	5.65
ET	35.5	-	104.0	79.6	77.0	13.7	438.8	116.0	328.3	122.3
NF	517	-	516	510	499	333	1056	520	929	594
NG	3102	-	1548	5109	4995	2671	13728	7283	4648	10707
NDF	117	-	101	104	108	74	244	118	199	138
NDG	702	-	304	1043	1082	594	3172	1661	998	2496
PNS	83.3	100.0	25.9	10.0	0	0	65.6	0	0	46.7
FFV	.47E-3	.74	.18E-7	.60E-2	-	-	.65E-8	-	-	.30
FVC	.47E+1	.12E-1	.15E+6	.0	-	-	.11E-3	-	-	.27E-3
F	0	1	1	0	0	0	2	0	1	0
PGS	80.0	0	80.0	0	0	10.0	75.0	0	0	12.5
GFV	-.22E+2	-	-.35E+3	-	-	-.17E-1	-.87E-1	-	-	-.16E-2
GVC	.24E-4	-	.94E-5	-	-	.41E-7	.10E-3	-	-	.55E-5

Table 63: Test results for FUNMIN (general problems).

Class	1B	2B	3B	4B	5B	6B	7B	8B	9B	10B
FV	.92E-8	.12E-9	.81E-9	.12E-11	.97E-10	.11E-9	.19E-6	.13E-10	.23E-10	.30E-10
VC	.77E-8	.50E-10	.67E-10	.13E-10	.20E-9	.11E-9	.43E-9	.18E-10	.12E-10	.21E-10
KT	.25E-4	.44E-5	.14E-4	.60E-7	.24E-5	.32E-5	.10E-3	.81E-7	.19E-6	.17E-6
ED	5.38	5.91	5.49	7.76	6.86	6.37	2.89	8.68	8.46	8.43
A	6.53	7.87	7.15	9.45	8.05	7.95	5.74	9.35	9.18	9.10
ET	72.9	82.6	73.2	23.1	61.6	93.4	118.8	28.6	28.0	27.3
NF	451	514	468	141	391	605	766	175	170	162
NG	4515	5142	4681	1412	3917	6058	7668	1751	1701	1628
NDF	99	111	101	32	83	129	162	40	39	37
NDG	992	1116	1010	321	838	1292	1625	403	395	372
ET/A	12.0	10.8	10.8	2.5	7.7	11.7	21.7	3.1	3.0	3.0
NF/A	74.5	67.3	69.4	15.4	48.9	75.6	139.7	18.9	18.5	18.0
NG/A	745.0	672.6	693.5	154.2	488.9	756.4	1396.7	188.7	184.6	179.8
NDF/A	16.4	14.6	15.0	3.5	10.5	16.1	29.8	4.3	4.3	4.1
NDG/A	164.1	146.1	150.2	35.1	104.7	160.9	297.5	43.4	42.8	41.0
R	0	0	0	0	0	0	0	0	0	0

Table 64: Test results for FUNMIN (degenerate, ill-conditioned, and indefinite problems).

Class	1A	2A	3A	4A	5A	6A	7A	8A	9A	10A
FV	.61E-4	-	.13E-5	.30E-5	.27E-4	.40E-6	-	.73E-3	.39E-4	.26E-3
VC	.45E-6	-	.68E-6	.24E-11	.16E-11	.42E-10	-	.0	.19E-8	.22E-7
KT	.60E-3	-	.16E-2	.13E-2	.45E-2	.89E-3	-	.69E-1	.73E-3	.27E-1
ED	4.66	-	3.71	4.27	3.28	3.98	-	2.67	3.01	3.12
ET	171.0	-	35.7	48.5	82.0	13.8	-	203.9	167.2	183.9
NF	344	-	58	105	163	131	-	291	122	332
NG	2067	-	174	1052	1632	1052	-	4074	611	5985
NDF	344	-	58	105	163	131	-	291	122	332
NDG	2067	-	174	1052	1632	1052	-	4074	611	5985
PNS	25.9	100.0	33.3	0	16.7	10.0	-	33.3	8.3	66.7
FFV	.20E-1	.97E-7	.20E-1	-	.74E-2	.72E-1	-	.75E-2	.52E-2	.52E-2
FVC	.72E-6	.19E+5	.16E-5	-	.0	.0	-	.78E-7	.11E-7	.13E-9
F	1	9	4	0	0	0	5	4	1	0
PGS	80.0	0	83.3	0	0	3.7	-	0	0	20.0
GFV	-.15E+1	-	-.43	-	-	-.17E-1	-	-	-	-.89E-3
GVC	.37E-6	-	.19E-5	-	-	.23E-7	-	-	-	.0

Table 65: Test results for GAPFPR (general problems).

Class	1B	2B	3B	4B	5B	6B	7B	8B	9B	10B
FV	.80E-4	.14E-4	.18E-4	.70E-4	.84E-4	.53E-4	.25E-3	.25E-2	.12E-2	.27E-2
VC	.97E-11	.24E-11	.19E-11	.37E-11	.0	.0	.0	.46E-10	.46E-10	.67E-10
KT	.76E-2	.32E-2	.41E-2	.92E-2	.80E-2	.21E-2	.28E-2	.18E-2	.61E-2	.43E-3
ED	2.78	2.99	3.02	2.67	3.21	3.06	2.02	3.24	3.50	3.35
A	5.00	5.49	5.47	5.07	5.34	5.50	5.05	4.73	4.74	4.86
ET	128.0	93.4	102.0	112.8	88.1	65.0	81.3	116.2	112.2	128.4
NF	243	181	195	215	172	123	154	214	208	238
NG	2435	1810	1952	2150	1724	1237	1545	2140	2080	2382
NDF	243	181	195	215	172	123	154	214	208	238
NDG	2435	1810	1952	2150	1724	1237	1545	2140	2080	2382
ET/A	27.4	16.9	19.0	23.5	17.0	11.9	16.2	25.6	23.3	26.4
NF/A	52.1	32.7	36.5	44.6	33.2	22.6	30.8	47.4	43.4	48.9
NG/A	521.5	327.2	364.5	446.1	332.3	226.1	307.6	474.3	432.5	488.7
NDF/A	52.1	32.7	36.5	44.6	33.2	22.6	30.8	47.4	43.4	48.9
NDG/A	521.5	327.2	364.5	446.1	332.3	226.1	307.6	474.3	432.5	488.7
R	0	0	0	0	1	1	0	3	1	2

Table 66: Test results for GAPFPR (degenerate, ill-conditioned, and indefinite problems).

Class	1A	2A	3A	4A	5A	6A	7A	8A	9A	10A
FV	.15E-4	.13E-5	.10E-5	.25E-5	.72E-6	.11E-5	-	.42E-6	.31E-5	.50E-6
VC	.55E-5	.49E-6	.23E-5	.63E-6	.14E-5	.24E-6	-	.72E-6	.11E-5	.49E-6
KT	.74E-2	.25E-5	.33E-5	.23E-3	.11E-3	.27E-5	-	.39E-5	.69E-4	.17E-5
ED	3.01	6.46	6.21	3.99	4.20	5.57	-	5.96	3.90	6.66
ET	43.4	77.6	28.8	66.4	69.9	10.6	-	73.3	214.3	61.4
NF	141	223	66	156	148	121	-	164	209	150
NG	1112	2217	377	1537	1515	912	-	2030	1443	1888
NDF	141	223	66	156	148	121	-	164	209	150
NDG	1112	2217	377	1537	1515	912	-	2030	1443	1888
PNS	0	52.4	50.0	0	6.7	6.7	-	0	0	0
FPV	-	.30E-4	.0	-	.61E-1	.83E-2	-	-	-	-
FVC	-	.37E-4	.59E+8	-	.90E-9	.55E-5	-	-	-	-
F	9	3	4	5	0	0	5	0	2	2
PGS	0	0	0	0	0	7.1	-	0	0	0
GFV	-	-	-	-	-	-.17E-1	-	-	-	-
GVC	-	-	-	-	-	.59E-9	-	-	-	-

Table 67: Test results for GAPFQL (general problems).

Class	1B	2B	3B	4B	5B	6B	7B	8B	9B	10B
FV	.29E-5	.31E-4	.30E-4	.15E-9	.79E-6	.10E-2	.13E-2	.12E-6	.37E-5	.18E-5
VC	.40E-5	.26E-7	.27E-7	.13E-5	.26E-8	.0	.92E+5	.12E-7	.37E-5	.11E-6
KT	.13E-2	.44E-2	.23E-2	.18E-5	.23E-3	.27E-1	.35E-1	.47E-6	.33E-5	.34E-5
ED	3.78	2.89	3.32	4.68	4.50	2.51	2.13	7.16	6.56	6.36
A	4.40	4.34	4.51	6.54	5.71	4.77	2.87	7.08	5.73	6.13
ET	121.2	103.0	114.0	13.5	43.5	185.2	131.4	40.9	77.1	77.3
NF	239	207	233	27	92	390	278	82	155	156
NG	2396	2076	2336	273	920	3900	2785	828	1558	1563
NDF	239	207	233	27	92	390	278	82	155	156
NDG	2396	2076	2336	273	920	3900	2785	828	1558	1563
ET/A	27.4	25.3	25.2	2.0	6.9	38.5	45.9	6.1	17.7	14.9
NF/A	54.4	51.5	51.9	4.1	14.5	80.8	97.3	12.3	35.7	30.3
NG/A	544.2	515.1	518.8	40.5	145.1	807.8	972.6	123.3	357.4	302.5
NDF/A	54.4	51.5	51.9	4.1	14.5	80.8	97.3	12.3	35.7	30.3
NDG/A	544.2	515.1	518.8	40.5	145.1	807.8	972.6	123.3	357.4	302.5
R	1	0	2	0	1	1	4	0	0	2

Table 68: Test result for GAPFQL (degenerate, ill-conditioned, and indefinite problems)

Class	1A	2A	3A	4A	5A	6A	7A	8A	9A	10A
FV	.16E-5	.11E-5	.16E-5	.39E-5	.37E-5	.36E-5	-	.72E-6	.16E-5	.11E-5
VC	.99E-6	.23E-5	.36E-5	.65E-8	.38E-7	.10E-6	-	.29E-5	.39E-5	.44E-5
KT	.36E-5	.32E-5	.15E-5	.30E-3	.60E-3	.15E-4	-	.28E-4	.66E-4	.56E-4
ED	6.16	7.93	6.48	4.08	3.83	5.06	-	5.82	3.65	5.59
ET	53.6	157.8	49.6	33.5	47.2	17.3	-	113.9	127.7	145.2
NF	446	614	79	109	148	153	-	225	132	343
NG	4126	10337	447	2003	2749	2396	-	5818	1243	10805
NDF	236	239	66	82	115	135	-	179	109	249
NDG	692	939	198	272	484	415	-	1408	347	2886
PNS	8.5	28.6	20.0	0	0	0	50.0	0	0	20.0
FFV	.51E-4	.84E+2	.0	-	-	-	.13	-	-	.24E-2
FVC	.15E+1	.18E-9	.48	-	-	-	.11E-5	-	-	.13E-5
F	1	3	0	0	0	0	3	0	1	0
PGS	59.1	6.7	87.5	0	0	10.0	100.0	0	0	0
GFV	-.69	-.20E+1	-.36E+1	-	-	-.17E-1	-.12E+1	-	-	-
GVC	.91E-5	.11E-5	.19E-5	-	-	.26E-8	.66E-7	-	-	-

Table 69: Test results for ACDPAC (general problems).

Class	1B	2B	3B	4B	5B	6B	7B	8B	9B	10B
FV	.26E-5	.17E-5	.13E-5	.25E-5	.40E-5	.16E-5	.12E-5	.14E-5	.16E-5	.76E-7
VC	.18E-5	.39E-8	.95E-6	.54E-10	.71E-9	.24E-7	.37E-8	.14E-9	.13E-6	.19E-7
KT	.15E-5	.92E-3	.51E-3	.15E-2	.13E-2	.46E-3	.20E-4	.12E-5	.29E-5	.80E-6
ED	4.72	3.46	3.78	3.23	4.02	3.79	2.60	6.24	6.24	6.49
A	5.46	5.17	4.75	5.48	5.36	5.13	5.42	6.96	6.11	6.86
ET	19.4	40.6	39.2	45.0	36.5	41.7	56.6	29.0	25.3	28.6
NF	57	125	130	147	116	130	189	89	80	101
NG	1033	2332	2317	2673	2122	2391	3381	1636	1445	1710
NDF	43	98	93	110	86	99	137	68	59	65
NDG	201	407	379	426	364	408	556	275	237	254
ET/A	3.7	8.1	8.3	8.2	7.0	8.2	10.6	4.2	4.2	4.3
NF/A	11.1	24.9	27.6	26.8	22.4	25.7	35.5	12.9	13.5	15.1
NG/A	200.0	462.7	490.7	488.8	404.9	469.9	633.1	236.2	242.3	255.9
NDF/A	8.3	19.6	19.7	20.3	16.3	19.5	25.8	9.9	9.9	9.8
NDG/A	38.9	81.0	80.6	78.4	69.1	79.6	103.8	40.1	39.1	37.6
R	0	0	0	0	0	0	0	0	0	0

Table 70: Test results for ACDPAC (degenerate, ill-conditioned, and indefinite problems).

Class	1A	2A	3A	4A	5A	6A	7A	8A	9A	10A
FV	.39E-4	.12E-3	.16E-7	.47E-5	.82E-5	.35E-5	.18E-4	.78E-3	.14E-3	-
VC	.39E-5	.48E-6	.41E-7	.55E-7	.19E-8	.87E-6	.51E-4	.0	.69E-5	-
KT	.30E-2	.37E-4	.36E-6	.76E-3	.15E-2	.68E-3	.17E-2	.38E-1	.72E-2	-
ED	3.72	6.44	7.63	3.70	3.33	3.87	3.47	2.68	2.67	-
ET	66.5	232.2	131.1	109.4	99.3	27.3	286.2	150.2	233.7	-
NF	1002	1516	652	663	596	643	665	694	602	-
NG	6828	19132	1956	6658	5972	5287	8883	9923	3024	-
NDF	136	141	127	166	165	163	163	162	161	-
NDG	567	1130	383	1569	1513	1259	2047	2132	761	-
PNS	36.7	50.0	70.0	0	0	3.3	53.3	66.7	20.0	100.0
FFV	.54E-6	.51E-1	.11E-10	-	-	.16E-3	.12E-10	.42E-7	.0	.71E-7
FVC	.42E-2	.70E-3	.26E+1	-	-	.53E-2	.45E-1	.60E-5	.59E+3	.13E-1
F	0	0	0	0	0	0	0	0	0	0
PGS	57.9	20.0	33.3	6.7	0	10.3	42.9	0	0	0
GFV	-.65E+1	-.20E+1	-.71E+1	-.12E-2	-	-.17E-1	-.38E-2	-	-	-
GVC	.71E-5	.15E-6	.41E-3	.19E-3	-	.30E-6	.12E-3	-	-	-

Table 71: Test results for FMIN(1) (general problems).

Class	1B	2B	3B	4B	5B	6B	7B	8B	9B	10B
FV	.51E-3	.58E-3	.63E-3	.41E-3	.44E-3	.48E-3	.53E-3	.68E-4	.83E-4	.83E-4
VC	.14E-9	.82E-11	.82E-11	.89E-11	.11E-10	.10E-10	.61E-11	.99E-8	.32E-6	.50E-6
KT	.35E-1	.32E-1	.31E-1	.33E-1	.23E-1	.20E-1	.21E-1	.33E-2	.36E-2	.30E-2
ED	2.51	2.52	2.51	2.47	2.85	2.70	2.41	4.80	4.68	4.79
A	4.28	4.58	4.57	4.60	4.71	4.67	4.65	4.86	4.42	4.42
ET	77.9	78.4	79.4	75.9	76.6	76.0	76.3	80.3	78.0	77.5
NF	456	460	464	461	471	468	471	483	475	477
NG	4570	4607	4650	4612	4714	4685	4719	4855	4767	4786
NDF	111	111	111	111	112	111	111	111	111	111
NDG	1090	1087	1087	1086	1094	1092	1094	1073	1065	1072
ET/A	19.0	17.4	17.6	16.8	16.6	16.6	16.7	17.1	18.1	18.0
NF/A	112.6	102.4	103.4	102.6	102.5	102.5	103.5	102.7	109.8	110.3
NG/A	1127.1	1024.1	1034.9	1026.6	1025.9	1025.4	1035.7	1030.6	1102.2	1107.4
NDF/A	27.3	24.8	24.9	24.8	24.4	24.5	24.4	23.8	25.8	25.9
NDG/A	265.7	239.5	239.9	238.9	235.5	236.4	237.9	229.5	247.9	249.8
R	0	0	0	0	0	0	0	0	0	0

Table 72: Test results for FMIN(1) (degenerate, ill-conditioned, and indefinite problems).

Class	1A	2A	3A	4A	5A	6A	7A	8A	9A	10A
FV	.37E-4	.52E-3	.10E-4	.91E-4	.18E-3	.54E-4	.44E-4	-	.67E-4	-
VC	.47E-4	.15E-3	.12E-3	.52E-6	.31E-7	.23E-6	.25E-3	-	.20E-5	-
KT	.49E-2	.76E-4	.50E-2	.32E-2	.63E-2	.69E-2	.88E-2	-	.45E-2	-
ED	3.59	5.88	3.16	2.83	2.61	2.83	2.45	-	2.69	-
ET	71.4	235.5	222.5	181.7	151.8	59.6	428.5	-	506.1	-
NF	1525	2187	2053	2027	2205	2434	2703	-	5020	-
NG	8061	21587	6160	19339	20410	18906	34964	-	23396	-
NDF	0	0	0	0	0	0	0	-	0	-
NDG	0	0	0	0	0	0	0	-	0	-
PNS	53.3	73.3	80.0	6.7	13.3	10.0	93.3	100.0	73.3	100.0
PFV	.11E-5	.33E-1	.81E-11	.40E-7	.91E-3	.16E-2	.51E-9	.21E-3	.79E-5	.84E-8
FVC	.46E-1	.49E-4	.15E+1	.40E-6	.70E-6	.30E-3	.16E-1	.29E-2	.23E-1	.47E-1
F	0	0	0	0	0	0	0	0	0	0
PGS	50.0	25.0	0	0	0	11.1	0	0	0	0
GFV	-.73E+1	-.20E+1	-	-	-	-.17E-1	-	-	-	-
GVC	.24E-3	.16E-7	-	-	-	.23E-9	-	-	-	-

Table 73: Test results for FMIN(2) (general problems).

Class	1B	2B	3B	4B	5B	6B	7B	8B	9B	10B
FV	.22E-4	.80E-5	.32E-4	.16E-4	.41E-3	.21E-3	.12E-3	.57E-3	.61E-3	.63E-3
VC	.30E-9	.19E-9	.23E-10	.35E-10	.53E-9	.28E-10	.11E-9	.50E-9	.48E-9	.30E-4
KT	.34E-2	.37E-2	.68E-2	.45E-2	.22E-1	.12E-1	.86E-2	.21E-1	.21E-1	.22E-1
ED	3.26	3.23	3.03	2.91	2.83	2.90	2.55	4.05	4.03	3.92
A	4.97	5.12	5.08	5.13	4.29	4.77	4.62	4.57	4.56	3.33
ET	282.1	253.3	205.1	197.1	116.9	139.1	140.7	127.0	127.2	124.8
NF	3350	3094	2474	2365	1409	1663	1660	1497	1506	1480
NG	33055	30432	24269	23196	13778	16326	16419	14474	14467	14312
NDF	0	0	0	0	0	0	0	0	0	0
NDG	0	0	0	0	0	0	0	0	0	0
ET/A	58.6	50.2	41.2	39.3	27.5	28.9	30.1	29.2	29.1	37.5
NF/A	695.3	613.3	496.9	471.9	331.8	346.5	355.8	344.9	346.4	445.4
NG/A	6875.4	6023.9	4888.9	4634.5	3242.1	3391.3	3506.3	3318.0	3306.2	4302.7
NDF/A	0.0	0.0	0.0	0.0	0.0	0.0	0.0	0.0	0.0	0.0
NDG/A	0.0	0.0	0.0	0.0	0.0	0.0	0.0	0.0	0.0	0.0
R	0	0	0	0	0	0	0	0	0	0

Table 74: Test results for FMIN(2) (degenerate, ill-conditioned, and indefinite problems).

Class	1A	2A	3A	4A	5A	6A	7A	8A	9A	10A
FV	.33E-5	.34E-5	.98E-7	.18E-7	.19E-7	.91E-8	-	.28E-6	-	.15E-3
VC	.97E-6	.52E-5	.22E-6	.10E-7	.29E-7	.20E-8	-	.52E-7	-	.29E-4
KT	.75E-3	.18E-5	.16E-5	.65E-4	.11E-3	.21E-5	-	.30E-3	-	.17E-1
ED	4.16	8.05	6.84	5.32	5.01	6.31	-	4.72	-	3.02
ET	54.1	145.8	206.8	178.9	217.2	44.2	-	530.2	-	260.9
NF	479	627	235	275	270	284	-	448	-	239
NG	3033	7715	705	2771	2730	2309	-	6421	-	4501
NDF	426	360	386	533	598	680	-	1169	-	1011
NDG	1723	3087	1159	5080	5559	5275	-	15550	-	17315
PNS	33.3	63.3	27.8	0	0	0	100.0	44.4	-	13.3
FFV	.23E-6	.58E-3	.0	-	-	-	.0	.42E-4	-	.29E-2
FVC	.25E+2	.62	.23E+1	-	-	-	.87	.85E-2	-	.99E-3
F	0	0	4	0	0	0	4	2	5	0
PGS	40.0	18.2	53.8	0	0	10.0	0	0	-	0
GFV	-.53E+1	-.20E+1	-.13E+2	-	-	-.17E-1	-	-	-	-
GVC	.11E-4	.38E-4	.88E-5	-	-	.18E-6	-	-	-	-

Table 75: Test results for PMIN(3) (genral problems).

Class	1B	2B	3B	4B	5B	6B	7B	8B	9B	10B
FV	.67E-7	.80E-7	.53E-7	.79E-7	.11E-6	.26E-6	.96E-8	.52E-8	.52E-8	.59E-8
VC	.38E-8	.13E-9	.78E-10	.0	.15E-7	.72E-8	.16E-8	.57E-9	.22E-9	.36E-9
KT	.14E-3	.29E-3	.26E-3	.30E-3	.48E-3	.13E-3	.24E-5	.83E-6	.56E-6	.76E-6
ED	4.82	4.22	4.36	3.96	4.51	4.56	3.89	7.78	7.92	7.73
A	6.06	6.19	6.33	6.65	5.64	5.79	6.58	7.85	8.03	7.88
ET	232.4	231.4	234.1	233.7	192.1	205.9	247.1	197.0	197.9	195.9
NF	228	234	238	228	198	219	252	203	202	205
NG	2309	2356	2407	2301	2005	2214	2536	2051	2037	2067
NDF	538	552	563	534	461	491	581	449	450	506
NDG	5305	5423	5524	5242	4516	4838	5769	4381	4377	4370
ET/A	38.3	37.4	36.9	35.2	33.2	35.1	37.8	25.6	25.0	25.3
NF/A	37.7	37.9	37.8	34.4	34.3	37.1	38.4	26.4	25.4	26.3
NG/A	381.0	381.9	381.3	346.5	346.1	374.8	386.4	265.9	256.2	265.4
NDF/A	88.9	89.6	89.1	80.4	79.5	83.5	88.9	58.0	56.7	57.4
NDG/A	875.0	876.8	872.6	789.7	779.8	822.1	882.9	565.1	548.4	559.9
R	0	0	0	0	0	0	0	0	0	0

Table 76: Test results for PMIN(3) (degenerate, ill-conditioned, and indefinite problems).

Class	1A	2A	3A	4A	5A	6A	7A	8A	9A	10A
PV	.22E-5	-	.18E-5	.74E-6	.11E-5	.19E-5	-	-	.38E-6	-
VC	.18E-5	-	.24E-5	.16E-5	.22E-5	.23E-5	-	-	.85E-6	-
KT	.26E-3	-	.22E-5	.13E-3	.17E-3	.15E-3	-	-	.25E-4	-
ED	5.03	-	6.47	4.43	4.49	4.46	-	-	3.43	-
ET	134.2	-	134.7	88.5	92.1	44.4	-	-	183.8	-
NF	2735	-	1844	694	784	1500	-	-	868	-
NG	16414	-	5532	6942	7846	12005	-	-	4343	-
NDF	212	-	106	82	94	163	-	-	76	-
NDG	1278	-	319	825	945	1306	-	-	381	-
PNS	46.7	-	44.4	3.3	0	0	-	-	0	-
FPV	.44E-5	-	.0	.17E-2	-	-	-	-	-	-
FVC	.37	-	.67E+3	.14E-5	-	-	-	-	-	-
F	0	10	4	0	0	0	5	5	2	5
PGS	81.3	-	80.0	0	0	6.7	-	-	0	-
GFV	-.85	-	-.13E+2	-	-	-.17E-1	-	-	-	-
GVC	.13E-5	-	.32E-5	-	-	.21E-5	-	-	-	-

Table 77: Test results for NLP (general problems).

Class	1B	2B	3B	4B	5B	6B	7B	8B	9B	10B
PV	.22E-5	.60E-6	.10E-5	.24E-9	.95E-6	.12E-5	.14E-5	.27E-5	.16E-5	.23E-5
VC	.23E-5	.14E-5	.17E-5	.20E-6	.21E-5	.24E-5	.28E-5	.36E-5	.35E-5	.35E-5
KT	.42E-3	.77E-4	.16E-3	.99E-5	.33E-3	.14E-3	.39E-4	.16E-4	.54E-5	.13E-4
ED	4.29	4.34	4.24	4.95	4.57	4.34	2.73	5.24	5.69	5.44
A	4.74	5.14	4.95	6.57	4.94	4.94	4.64	5.27	5.56	5.36
ET	112.3	81.9	86.7	37.3	92.6	111.2	143.9	180.6	177.4	170.3
NF	1025	732	770	286	827	1030	1392	1758	1720	1588
NG	10252	7321	7707	2860	8276	10305	13920	17585	17208	15881
NDF	100	76	83	43	86	99	112	129	133	126
NDG	1005	760	831	432	866	990	1121	1292	1336	1266
ET/A	23.6	16.1	17.6	6.1	18.7	22.6	31.0	34.9	32.4	32.6
NF/A	215.4	143.8	156.3	46.9	167.1	209.4	299.8	341.9	314.8	306.3
NG/A	2154.4	1437.9	1562.6	469.2	1670.5	2093.5	2998.2	3419.2	3147.6	3062.6
NDF/A	21.2	14.9	16.8	7.0	17.5	20.2	24.2	25.0	24.4	24.1
NDG/A	212.0	148.5	168.3	69.9	175.3	201.7	241.8	249.8	243.5	241.4
R	0	0	0	0	0	0	0	0	0	0

Table 78: Test results for NLP (degenerate, ill-conditioned, and indefinite problems).

Class	1A	2A	3A	4A	5A	6A	7A	8A	9A	10A
FV	.47E-4	-	.29E-3	.27E-5	.27E-6	.56E-3	.40E-4	.11E-5	.34E-5	.26E-5
VC	.13E-5	-	.42E-3	.0	.0	.0	.31E-5	.0	.28E-6	.0
KT	.25E-2	-	.53E-2	.11E-2	.33E-3	.31E-1	.18E-2	.49E-3	.10E-2	.13E-2
ED	3.72	-	3.37	3.93	4.52	2.28	3.41	4.62	3.17	4.33
ET	107.3	-	183.2	211.5	232.4	65.7	302.3	361.8	453.2	602.7
NF	1740	-	1474	1915	2513	2125	1391	2535	2909	3963
NG	11750	-	4419	20540	26345	18983	18555	37978	15249	77788
NDF	75	-	56	79	102	95	70	130	117	208
NDG	498	-	168	859	1063	840	921	1918	603	4014
PNS	79.2	100.0	62.5	63.3	50.0	66.7	93.3	66.7	60.0	66.7
FFV	.16E-5	.14E+3	.12E-9	.11	.42E-1	.18	.47E-6	.24	.34E-3	.18E+1
FVC	.20E+1	.0	.84E-2	.33E-9	.76E-10	.34E-6	.55E-1	.0	.31E-2	.0
F	2	3	2	0	0	8	0	1	0	3
PNG	40.0	0	66.7	0	0	50.0	0	0	0	0
GFV	-.12E+1	-	-.16	-	-	-.17E-1	-	-	-	-
GVC	.10E-3	-	.22E-3	-	-	.0	-	-	-	-

Table 79: Test results for SUMT (general problems).

Class	1B	2B	3B	4B	5B	6B	7B	8B	9B	10B
FV	.32E-4	.31E-4	.31E-4	.16E-4	.37E-5	.26E-4	.41E-4	.22E-4	.25E-4	.21E-4
VC	.0	.0	.0	.0	.0	.0	.0	.0	.0	.0
KT	.34E-2	.50E-2	.37E-2	.48E-2	.11E-2	.32E-2	.11E-2	.75E-4	.89E-4	.25E-4
ED	3.30	3.05	3.08	2.95	4.05	3.26	2.69	5.06	5.09	5.13
A	5.57	5.47	5.51	5.51	6.11	5.58	5.51	6.46	6.43	6.60
ET	101.5	101.0	101.0	101.4	100.6	101.8	101.4	96.5	97.9	99.4
NF	1019	1042	1030	1036	1042	1052	1029	918	938	950
NG	10398	10608	10495	10559	10659	10725	10490	9965	10188	10314
NDF	53	54	54	54	51	54	54	48	48	49
NDG	536	550	552	550	520	546	548	516	508	522
ET/A	18.3	18.5	18.3	18.4	16.5	18.2	18.4	15.1	15.3	15.3
NF/A	183.3	190.8	187.2	188.4	170.6	188.5	187.0	143.6	147.2	146.1
NG/A	1869.5	1941.8	1906.3	1918.3	1744.7	1921.2	1905.1	1556.7	1595.5	1583.3
NDF/A	9.6	10.0	9.9	9.9	8.4	9.7	9.9	7.7	7.5	7.6
NDG/A	96.5	100.6	100.4	99.8	85.1	97.8	99.6	80.9	79.7	80.4
R	0	0	0	0	0	0	0	0	0	0

Table 80: Test results for SUMT (degenerate, ill-conditioned, and indefinite problems).

Class	1A	2A	3A	4A	5A	6A	7A	8A	9A	10A
FV	.36E-6	.25E-3	.97E-8	.28E-6	.20E-8	-	-	.57E-7	.37E-4	-
VC	.15E-6	.50E-3	.21E-7	.37E-10	.37E-9	-	-	.35E-10	.23E-3	-
KT	.72E-4	.88E-5	.52E-5	.44E-3	.57E-4	-	-	.17E-3	.20E-2	-
ED	5.37	6.78	6.56	4.66	5.67	-	-	5.08	2.76	-
ET	83.1	225.1	84.0	50.1	54.6	-	-	242.7	184.0	-
NF	1172	1355	422	334	382	-	-	1258	582	-
NG	7034	16265	1266	3344	3822	-	-	17613	2913	-
NDF	170	160	72	50	57	-	-	173	86	-
NDG	1025	1921	217	507	573	-	-	2429	431	-
PNS	14.3	42.9	4.8	29.2	56.7	-	-	26.7	50.0	-
FFV	.28E+1	.60E-5	.65E-2	.22E+33	.19E-1	-	-	.23E+10	.28E+18	-
FVC	.17E+1	.53E+1	.43E-6	.59E+14	.41E-6	-	-	.11E+6	.98E+25	-
F	3	3	3	2	0	10	5	0	3	5
PGS	55.6	0	75.0	0	0	-	-	0	0	-
GFV	-.99	-	-.31E+1	-	-	-	-	-	-	-
GVC	.30E-6	-	.18E-4	-	-	-	-	-	-	-

Table 81: Test results for DFP (general problems).

Class	1B	2B	3B	4B	5B	6B	7B	8B	9B	10B
FV	.90E-4	.14E-4	.16E-3	.17E-9	.10E-3	.23E-2	.57E-3	.94E-8	.15E-8	.30E-8
VC	.10E-7	.11E-3	.13E-8	.34E-6	.0	.34E-12	.68E-3	.68E-11	.15E-10	.13E-10
KT	.14E-1	36E-2	.13E-1	.84E-6	.12E-1	.59E-1	.37E-2	.37E-4	.24E-4	.14E-4
ED	2.98	3.11	2.64	4.84	3.38	2.34	1.97	7.18	7.79	7.75
A	4.21	3.58	4.30	6.79	5.08	4.67	2.70	7.70	8.02	8.01
ET	68.3	51.4	118.6	38.2	36.8	112.9	199.4	42.9	47.7	43.2
NF	510	362	899	259	281	954	1755	297	329	297
NG	5100	3624	8994	2592	2810	9540	17550	2971	3293	2977
NDF	66	52	117	44	40	109	184	45	49	43
NDG	664	528	1172	440	400	1090	1840	453	491	435
ET/A	17.3	15.4	25.3	6.0	7.2	24.2	73.9	5.6	6.0	5.4
NF/A	129.8	108.6	190.7	40.5	55.3	204.3	650.0	38.7	41.4	37.5
NG/A	1297.5	1086.0	1907.1	404.9	553.1	2042.8	6500.0	387.0	414.3	374.9
NDF/A	16.9	15.9	25.0	6.8	7.9	23.3	68.1	5.9	6.2	5.5
NDG/A	169.4	159.0	249.9	68.4	78.7	233.4	681.4	59.2	61.8	54.6
R	1	2	2	0	1	2	7	0	0	0

Table 82: Test results for DFP (degenerate, ill-conditioned, and indefinite problems)

Class	1A	2A	3A	4A	5A	6A	7A	8A	9A	10 A
FV	.65E-4	.23E-5	.23E-3	.11E-5	.34E-5	.72E-7	.96E-4	.37E-4	.18E-4	.59E-5
VC	.95E-4	.59E-8	.28E-3	.10E-11	.0	.48E-9	.20E-3	.35E-10	.64E-4	.74E-5
KT	.38E-4	.15E-6	.44E-4	.33E-3	.64E-3	.48E-4	.11E-2	.27E-2	.15E-3	.89E-3
ED	5.30	9.31	5.35	4.38	4.05	5.59	3.36	3.87	3.34	4.90
ET	16.2	11.0	31.1	9.3	14.5	13.3	100.9	18.3	68.5	45.6
NF	208	57	76	78	107	208	199	61	159	160
NG	429	353	204	195	306	513	1234	438	263	1657
NDF	94	28	43	40	53	107	105	30	81	85
NDG	224	183	117	97	147	276	669	217	139	905
PNS	56.7	66.7	20.0	20.0	6.7	16.7	13.3	0	0	0
FFV	.33E-5	.21E-7	.30	.32E+1	.25E+1	.84	.81E-6	-	-	-
FVC	.16E+2	.30E+3	.27E-1	.15E-1	.10E-1	.25	.77E-1	-	-	-
F	0	1	0	0	0	0	0	0	1	0
PGS	69.2	22.2	95.8	0	0	12.0	84.6	0	0	20.0
GFV	-.22E+2	-.20E+1	-.65E+3	-	-	-.17E-1	-.51E-1	-	-	-.17E-2
GVC	.12E-3	.46E-9	.30E-3	-	-	.23E-9	.21E-3	-	-	.63E-4

Table 83: Test results for PCDPAK (general problems).

Class	1B	2B	3B	4B	5B	6B	7B	8B	9B	10B
FV	.52E-5	.20E-5	.27E-4	.84E-6	.30E-5	.20E-4	.27E-4	.15E-4	.15E-4	.15E-4
VC	.0	.0	.0	.0	.0	.0	.0	.0	.0	.0
KT	.14E-2	.64E-3	.29E-2	.83E-3	.13E-2	.20E-2	.20E-3	.17E-3	.18E-3	.13E-3
ED	3.71	3.88	3.18	3.48	4.03	3.35	2.65	4.85	4.90	4.88
A	5.96	6.19	5.57	6.16	6.11	5.69	5.73	6.36	6.36	6.40
ET	6.3	5.3	5.0	4.2	5.4	4.4	4.3	3.1	3.0	3.1
NF	44	36	39	26	36	32	23	13	12	12
NG	116	88	75	66	94	64	70	49	50	50
NDF	22	19	19	13	18	16	11	8	7	7
NDG	54	42	34	30	44	29	33	24	25	26
ET/A	1.1	.9	.9	.7	.9	.8	.8	.5	.5	.5
NF/A	7.5	6.0	7.2	4.2	6.0	5.7	4.1	2.2	2.0	2.0
NG/A	19.4	14.1	13.8	10.5	15.3	11.4	12.3	7.8	7.9	7.9
NDF/A	3.8	2.9	3.5	2.1	3.0	2.9	2.1	1.3	1.2	1.2
NDG/A	9.1	6.7	6.3	4.9	7.1	5.3	5.8	3.9	4.0	4.1
R	0	0	0	0	0	0	0	0	0	0

Table 84: Test results for PCDPAK (degenerate, ill-conditioned, and indefinite problems).

Code	A	ET	NF	NG	NDF	NDG	R
OPRQP	8.14	5.7	28	221	27	208	0
XROP	7.65	2.9	16	127	16	127	0
VF02AD	6.59	27.4	10	82	10	82	0
GRGA	4.96	7.7	89	996	37	132	1
OPT	4.85	18.9	574	4669	0	143	1
GRG2(1)	3.47	15.1	258	1851	40	297	4
GRG2(2)	3.35	45.4	1275	13244	0	0	4
VF01A	7.40	9.2	91	686	91	309	0
LPNLP	6.33	11.2	185	1247	54	368	0
SALQDR	3.86	23.0	89	741	88	731	0
SALQDF	4.26	33.4	1082	8351	0	0	0
SALMNF	8.54	49.3	32	1682	212	1682	0
CONMIN	5.67	15.4	961	5175	76	413	2
BIAS(1)	5.43	13.8	338	2882	45	333	0
BIAS(2)	3.21	36.5	1576	11517	0	0	0
FUNMIN	8.06	5.9	236	1068	54	248	6
GAPFPR	4.47	31.7	169	1340	169	1340	2
GAPFQL	6.10	13.3	101	685	101	685	0
ACDPAC	5.42	21.5	124	1748	94	421	0
FMIN(1)	4.17	33.9	666	5109	152	1019	0
FMIN(2)	3.86	65.7	2778	17731	0	0	0
FMIN(3)	6.94	105.4	300	2206	608	4470	0
NLP	5.29	86.6	2405	18292	168	1277	0
SUMT	5.97	82.5	2168	23271	95	996	18
DFP	7.17	13.2	412	2265	59	326	7
FCDPAK	5.67	5.3	35	144	21	89	0

Table 85: Test results for convex problems (starting point close to the solution).

Code	A	ET	NF	NG	NDF	NDG	R
OPRQP	8.08	13.2	67	598	50	423	6
XROP	8.02	9.3	47	451	36	323	8
VF02AD	6.70	69.6	25	213	25	213	1
GRGA	6.79	53.3	494	8709	159	620	1
OPT	4.73	58.8	1170	14496	0	373	10
GRG2(1)	4.08	34.5	780	5920	67	512	8
GRG2(2)	3.70	49.2	1749	14081	0	0	6
VF01A	7.24	23.2	159	1159	159	1159	11
LPNLP	6.41	12.1	196	1378	61	444	10
SALQDR	4.73	52.8	179	1549	178	1540	0
SALQDF	4.50	56.3	1736	14130	0	0	0
SALMNF	8.48	91.0	68	3157	355	3157	0
CONMIN	3.98	19.2	1231	4981	83	347	20
BIAS(1)	5.11	31.8	1035	7158	84	591	11
BIAS(2)	3.93	59.2	2780	18949	0	0	8
FUNMIN	6.09	33.7	1362	6310	311	1442	20
GAPFPR	4.82	23.3	150	1027	150	1027	8
GAPFQL	6.13	58.9	214	1843	214	1843	5
ACDPAC	5.21	34.1	209	2851	144	605	0
FMIN(1)	3.44	39.4	1069	8605	143	648	8
FMIN(2)	2.76	60.6	2856	15753	0	0	6
FMIN(3)	4.96	61.1	300	2235	406	2153	5
NLP	5.34	132.7	3206	26987	253	2147	7
SUMT	5.77	103.2	2809	29098	124	1284	20
DFP	6.83	27.2	800	4588	115	664	10
FCDPAK	6.87	67.1	499	2347	226	1103	7

Table 86: Test results for convex problems (starting point far away from the solution).

EVALUATION OF SIGNIFICANCE FACTORS

The evaluation of the ease of use items of Table 15 and the weight factors of Tables 7, 14, and 18 is based on the priority theory of Saaty [SY 1975, SY 1977] which was applied to the utilization of the results of former comparative studies by Lootsma [LO 1979]. To give a rough sketch of the theory, consider n factors $F_1,\ldots,F_n$, and assume that they have positive priorities $w_1,\ldots,w_n$ with

$$\sum_{i=1}^{n} w_i = 1.$$

The matrix

$$A := (w_i/w_j)_{i,j=1,n}$$

is of rank one and because of

$$Aw = nw ,$$

A has only one non-zero eigenvalue n with the corresponding eigen-vector $w := (w_1,\ldots,w_n)^T$. A further property of A is that the row sums

$$w_i \sum_{j=1}^{n} \frac{1}{w_j}$$

and the inverse column sums

$$(\frac{1}{w_j} \sum_{i=1}^{n} w_i)^{-1}$$

are always multiples of the vector w.

It is now assumed that it is impossible to obtain the priorities $w_1,\ldots,w_n$ by direct measurement, but that we are able to determine at least relative significances r_{ij} for each pair of factors F_i and F_j indicating whether F_i is more, less, or equally important than F_j. This leads to a matrix

$$R := (r_{ij})_{i,j=1,n}$$

which is considered as a perturbation of the matrix A. Since we require that $r_{ii} = 1$ and $r_{ij} \cdot r_{ji} = 1$, R is reciprocal with positive elements. For an approximation of the true priorities $w_1, \ldots, w_n$, we compute the normalized row sums

$$\sum_{j=1}^{n} r_{ij} / \sum_{k=1}^{n} \sum_{j=1}^{n} r_{kj}$$

or the normalized inverse column sums

$$(\sum_{k=1}^{n} r_{kj})^{-1} / \sum_{i=1}^{n} (\sum_{k=1}^{n} r_{ki})^{-1}.$$

For applying priority theory to our performance evaluation of optimization programs, we define the following relative significances:

$r_{ij} = 1$, if F_i and F_j are equally significant.

$r_{ij} = 3$, if F_i is somewhat more important than F_j.

$r_{ij} = 5$, if F_i is much more important than F_j.

The intermediate values 2 and 4 could be used in cases of doubt and higher ones for describing extraordinary significances.

First, we evaluate scores for the ease of use items quality of documentation, provision of problem data and functions, program organization, and sensitivity to input parameters. The results are presented in Tables 87 to 90. The columns headed by NRS show the normalized row sums and the rows identified by NICS the normalized inverse column sums. The mean value of two corresponding NRS and NICS values defines the score for the performance of a program as reported in Table 15.

In a similar way, the weight factors of Tables 7 and 14 for evaluating the performance criteria efficiency (E1, E2), reliability (R), global convergence (G), and ease of use (EU) were determined. We have to point out, however, that the relative significances as given in Tables 91 to 95 should be considered only as proposals and a decision maker is free to define his own significances when selecting a program.

In Section 7 of Chapter V we outlined how one could get final scores for selecting optimization programs. Again, one has to define weights for each of the nine performance criteria, and these weights may also be obtained using priority theory, cf. Tables 96 to 98.

Code		C1	C2	C3	C4	C5	C6	C7	C8	C9	C10	C11	C12	C13	C14	C15	C16	C17	C18	C19	NRS
OPRQP/XROP	C1	1	1	1	1/3	1/3	1	1/3	1/3	3	1/3	4	4	3	1/3	4	1/3	3	2	1/3	.051
VFO2AD	C2	1	1	1	1/3	1/3	1	1/3	1/3	3	1/3	4	4	3	1/3	4	1/3	3	2	1/3	.051
GRGA	C3	1	1	1	1/3	1/4	1	1/3	1/4	2	1/3	3	3	3	1/3	3	1/3	2	1	1/3	.041
OPT	C4	3	3	3	1	1	3	1	1/2	3	1	4	4	3	1	4	1	3	3	1	.075
GRG2(1/2)	C5	3	3	4	1	1	2	1/2	1/2	3	1	4	4	4	1	4	1	3	3	1	.076
VFO1A	C6	1	1	1	1/3	1/2	1	1/3	1/3	3	1/3	4	4	4	1/3	4	1/3	3	2	1/3	.053
LPNLP	C7	3	3	3	1	2	3	1	1	3	2	5	5	4	2	5	2	3	3	2	.091
SALQDR/-QDF/-MNF	C8	3	3	4	2	2	3	1	1	4	2	5	5	5	2	5	2	4	3	2	.100
CONMIN	C9	1/3	1/3	1/2	1/3	1/3	1/3	1/3	1/4	1	1/3	3	3	3	1/3	3	1/4	1	1	1/3	.033
BIAS(1/2)	C10	3	3	3	1	1	3	1/2	1/2	3	1	4	4	4	1	4	1	3	2	1	.074
FUNMIN	C11	1/4	1/4	1/3	1/4	1/4	1/4	1/5	1/5	1/3	1/4	1	1/2	1/3	1/4	1	1/4	1/3	1/3	1/4	.012
GAPFPR	C12	1/4	1/4	1/3	1/4	1/4	1/4	1/5	1/5	1/3	1/4	2	1	1/2	1/4	1	1/4	1/3	1/3	1/4	.015
GAPFQL	C13	1/3	1/3	1/3	1/3	1/4	1/4	1/4	1/5	1/3	1/4	3	2	1	1/4	2	1/4	1/3	1/2	1/4	.021
ACDPAC	C14	3	3	3	1	1	3	1/2	1/2	3	1	4	4	4	1	4	1	3	2	1	.074
FMIN(1/2/3)	C15	1/4	1/4	1/3	1/4	1/4	1/4	1/5	1/5	1/3	1/4	1	1	1/2	1/4	1	1/4	1/3	1/3	1/4	.013
NLP	C16	3	3	3	1	1	3	1/2	1/2	4	1	4	4	4	1	4	1	4	3	1	.079
SUMT	C17	1/3	1/3	1/2	1/3	1/3	1/3	1/3	1/4	1	1/3	3	3	3	1/3	3	1/4	1	1	1/3	.033
DFP	C18	1/2	1/2	1	1/3	1/3	1/2	1/3	1/3	1	1/2	3	3	2	1/2	3	1/3	1	1	1/2	.034
FCDPAK	C19	3	3	3	1	1	3	1/2	1/2	3	1	4	4	4	1	4	1	3	2	1	.074
NICS·10^{-3}		34	34	31	83	77	35	119	131	25	76	16	16	19	75	16	78	25	32	76	

Table 87: Quality of documentation

Code		C1	C2	C3	C4	C5	C6	C7	C8	C9	C10	C11	C12	C13	C14	C15	C16	C17	C18	C19	NRS
OPRQP/XROP	C1	1	1/3	1/5	3	1	1/2	1	1/5	5	3	3	4	1/4	3	1	3	5	5	4	.066
VF02AD	C2	3	1	1/3	5	2	2	3	1/3	5	5	5	5	1/3	5	3	4	5	5	5	.098
GRGA	C3	5	3	1	5	3	3	4	1	5	5	5	5	1	5	5	5	5	5	5	.116
OPT	C4	1/3	1/5	1/5	1	1/2	1/3	1/2	1/5	3	1	1	2	1/4	1	1/2	3	3	3	2	.035
GRG2(1/2)	C5	1	1/2	1/3	2	1	1	2	1/5	5	3	3	3	1/3	2	2	4	5	5	3	.066
VF01A	C6	2	1/2	1/3	3	1	1	1	1/3	5	3	3	4	1/3	3	3	4	5	5	3	.073
LPNLP	C7	1	1/3	1/4	2	1/2	1	1	1/5	5	3	3	3	1/4	2	2	4	5	5	3	.063
SALQDR/-QDF /-MNF	C8	5	3	1	5	3	3	5	1	5	5	5	5	2	5	5	5	5	5	5	.119
CONMIN	C9	1/5	1/5	1/5	1/3	1/5	1/5	1/5	1/5	1	1/5	1/5	1/3	1/5	3	1/5	1/3	1/3	1/3	1/2	.009
BIAS(1/2)	C10	1/3	1/5	1/5	1	1/3	1/3	1/3	1/5	5	1	1	2	1/4	1	1	2	3	3	2	.037
FUNMIN	C11	1/3	1/5	1/5	1	1/3	1/3	1/3	1/5	5	1	1	2	1/4	1	1	2	3	3	2	.037
GAPFPR	C12	1/4	1/5	1/5	1/2	1/3	1/4	1/3	1/5	3	1/2	1	1	1/4	1/2	1/2	1	2	2	1	.024
GAPFQL	C13	4	3	1	4	3	3	4	1/2	5	4	4	4	1	4	4	4	5	4	4	.100
ACDPAC	C14	1/3	1/5	1/5	1	1/2	1/3	1/2	1/5	3	1	1	2	1/4	1	1/2	3	3	3	2	.035
FMIN(1/2/3)	C15	1	1/3	1/5	2	1/2	1/3	1/2	1/5	5	1	1	2	1/4	2	1	3	4	4	3	.046
NLP	C16	1/3	1/4	1/5	1/3	1/4	1/4	1/4	1/5	3	1/2	1/2	1	1/4	1/3	1/3	1	3	3	1/2	.024
SUMT	C17	1/5	1/5	1/5	1/3	1/5	1/5	1/5	1/5	3	1/3	1/3	1/2	1/5	1/3	1/4	1/3	1	1	1/2	.015
DFP	C18	1/5	1/5	1/5	1/3	1/5	1/5	1/5	1/5	3	1/3	1/3	1/2	1/5	1/3	1/4	1/3	1	1	1/2	.015
FCDPAK	C19	1/4	1/5	1/5	1/2	1/3	1/3	1/3	1/5	2	1/2	1/2	1	1/4	1/2	1/3	2	2	2	1	.022
NICS·10⁻³		41	76	160	28	58	60	43	174	14	28	28	23	130	28	33	21	16	16	23	

Table 88: Provision of problem data and functions.

Code		C1	C2	C3	C4	C5	C6	C7	C8	C9	C10	C11	C12	C13	C14	C15	C16	C17	C18	C19	NRS
OPRQP/XROP	C1	1	1/5	1	1/3	1/2	1/3	1/2	1/5	1/2	1/4	1/2	1	1/2	1/2	1/4	1/3	1/2	1	1/2	.019
VF02AD	C2	5	1	5	3	4	3	4	1	4	2	4	5	5	5	2	3	4	5	5	.132
GRGA	C3	1	1/5	1	1/3	1/2	1/3	1/2	1/5	1/2	1/4	1/2	1	1	1	1/4	1/3	1/2	1	1	.021
OPT	C4	3	1/3	3	1	2	1	2	1/3	2	1/2	2	3	3	3	1/2	1	2	3	3	.067
GRG2(1/2)	C5	2	1/4	2	1/2	1	1/2	1	1/4	1	1/3	1	2	2	2	1/3	1/2	1	2	2	.041
VF01A	C6	3	1/3	3	1	2	1	2	1/3	2	1/2	2	3	3	3	1/2	1	2	3	3	.067
LPNLP	C7	2	1/4	2	1/2	1	1/2	1	1/4	1	1/3	1	2	2	2	1/3	1/2	1	2	2	.041
SALQDR/-QDF /-MNF	C8	5	1	5	3	4	3	4	1	4	2	4	5	5	5	2	3	4	5	5	.132
CONMIN	C9	2	1/4	2	1/2	1	1/2	1	1/4	1	1/3	1	2	1	1	1/3	1/2	2	2	1	.035
BIAS(1/2)	C10	4	1/2	4	2	3	2	3	1/2	3	1	3	4	3	3	1	2	3	4	3	.092
FUNMIN	C11	2	1/4	2	1/2	1	1/2	1	1/4	1	1/3	1	2	1	1	1/3	1/2	2	2	1	.035
GAPFPR	C12	1	1/5	1	1/3	1/2	1/3	1/2	1/5	1/2	1/4	1	1	1/2	1/2	1/4	1/3	1/2	1	1/2	.019
GAPFQL	C13	2	1/5	1	1/3	1/2	1/3	1/2	1/5	1/3	1/3	2	2	1	1	1/3	1/2	1/2	1	1	.028
ACDPAK	C14	2	1/5	1	1/3	1/2	1/3	1/2	1/5	1/3	1/3	2	2	1	1	1/3	1/2	1/2	1	1	.028
FMIN(1/2/3)	C15	4	1/2	4	2	3	2	3	1/2	3	1	3	4	3	3	1	2	3	4	3	.092
NLP	C16	3	1/3	3	1	2	1	2	1/3	2	1/2	2	3	2	2	1/2	1	2	3	2	.062
SUMT	C17	2	1/4	2	1/2	1	1/2	1	1/4	1	1/3	1	2	2	2	1/3	1/2	1	2	2	.041
DFP	C18	1	1/5	1	1/3	1/2	1/3	1/2	1/4	1/2	1/4	1/2	1	1	1	1/4	1/3	1/2	1	1	.021
FCDPAK	C19	2	1/5	1	1/3	1/2	1/3	1/2	1/5	1/3	1/3	2	2	1	1	1/3	1/2	1/2	1	1	.028
NICS·10^{-3}		22	153	23	57	36	57	36	153	34	91	34	22	27	27	91	55	36	23	27	

Table 89: Program organization.

Code		C1	C2	C3	C4	C5	C6	C7	C8	C9	C10	C11	C12	C13	C14	C15	C16	C17	C18	C19	NRS
OPRQP/XROP	C1	1	1/2	1/2	3	1	1	3	1	5	3	4	3	1	1/2	5	2	5	4	1/2	.090
VFO2AD	C2	2	1	1	3	1	2	3	2	5	3	4	3	2	1	5	3	5	4	1	.085
GRGA	C3	2	1	1	3	2	2	3	2	5	3	5	3	2	1	5	3	5	5	1	.090
OPT	C4	1/3	1/3	1/3	1	1/3	1/3	1	1/3	3	1	2	1	1/2	1/3	3	1/2	4	2	1/3	.036
GRG2(1/2)	C5	1	1	1/2	3	1	1	3	1	5	3	4	3	1	1/2	5	2	5	4	1/2	.074
VFO1A	C6	1	1/2	1/2	3	1	1	3	1	5	3	3	3	1	1/2	5	2	5	3	1/2	.070
LPNLP	C7	1/3	1/3	1/3	1	1/3	1/3	1	1/3	3	1	2	1	1/2	1/3	3	1/2	4	2	1/3	.036
SALQDR/-QDF /-MNF	C8	1	1/2	1/2	3	1	1	3	1	5	3	3	3	1	1/2	4	2	5	3	1/2	.068
CONMIN	C9	1/5	1/5	1/5	1/3	1/5	1/5	1/3	1/5	1	1/3	1/3	1/3	1/4	1/5	1	1/4	2	1/2	1/5	.014
BIAS(1/2)	C10	1/3	1/3	1/3	1	1/3	1/3	1	1/3	3	1	2	1	1/2	1/3	3	1/2	4	2	1/3	.036
FUNMIN	C11	1/4	1/4	1/5	1/2	1/4	1/3	1/2	1/3	3	1/2	1	2	1/3	1/5	2	1/3	3	1	1/5	.024
GAPFPR	C12	1/3	1/3	1/3	1	1/3	1/3	1	1/3	3	1	2	1	1/2	1/3	3	1/2	4	2	1/3	.036
GAPFQL	C13	1	1/2	1/2	2	1	1	2	1	4	2	2	2	1	1/2	4	2	4	3	1/2	.058
ACDPAC	C14	2	1	1	3	2	2	3	2	5	3	5	3	2	1	5	3	5	5	1	.090
FMIN(1/2/3)	C15	1/5	1/5	1/3	1/5	1/5	1/5	1/3	1/4	1	1/3	1/2	1/3	1/4	1/5	1	1/4	3	1/2	1/5	.016
NLP	C16	1/2	1/3	1/3	2	1/2	1/2	2	1/2	4	2	3	2	1/2	1/3	4	1	5	3	1/3	.053
SUMT	C17	1/5	1/5	1/5	1/5	1/5	1/5	1/4	1/5	1/2	1/4	1/3	1/4	1/4	1/5	1/3	1/5	1	1/3	1/5	.009
DFP	C18	1/4	1/4	1/5	1/2	1/4	1/3	1/2	1/3	2	1/2	1	1/2	1/3	1	2	1/3	3	1	1/3	.023
FCDPAK	C19	2	1	1	3	2	2	3	2	5	3	5	3	2	1	5	3	5	5	1	.090
NICS·10^{-3}		65	105	111	30	68	63	30	63	15	30	20	30	60	111	16	39	13	17	111	

Table 90: Sensitivity to input parameters.

The abbreviations are explained in Table 18 where the final weight factors are listed. The relative significances of these tables represent situations where efficiency, ease of use, or reliability are most important, but a decision maker could choose other significances according to the individual application of the desired code.

Item	ET	NF	NG	NDF	NDG	NRS
ET	1	5	5	3	3	.44
NF	1/5	1	1	1/3	1/3	.07
NG	1/5	1	1	1/3	1/3	.07
NDF	1/3	3	3	1	1	.21
NDG	1/3	3	3	1	1	.21
NICS	.48	.08	.08	.18	.18	1.00

Table 91: Relative significance of efficiency items
(execution time and number of function/gradient
evaluations equally important).

Item	ET	NF	NG	NDF	NDG	NRS
ET	1	1/3	1/3	1/5	1/5	.06
NF	3	1	1	1/3	1/3	.14
NG	3	1	1	1/3	1/3	.14
NDF	5	3	3	1	1	.33
NDG	5	3	3	1	1	.33
NICS	.06	.12	.12	.35	.35	1.00

Table 92: Relative significance of efficiency items
(execution time less important than number
of function/gradient evaluations).

Item	PNS	FFV	FVC	F	NRS
PNS	1	4	3	3	.50
FFV	1/4	1	1/2	1/2	.10
FVC	1/3	2	1	1	.20
F	1/3	2	1	1	.20
NICS	.52	.12	.18	.18	1.00

Table 93: Relative significance of reliability items.

Item	PGS	GFV	GVC	NRS
PGS	1	3	5	.59
GFV	1/3	1	3	.29
GVC	1/5	1/3	1	.12
NICS	.61	.26	.13	1.00

Table 94: Relative significance of global convergence items.

Item	QD	PD	PO	SI	NRS
QD	1	2	3	1	.37
PD	1/2	1	2	1	.24
PO	1/3	1/2	1	1/2	.12
SI	1	1	2	1	.27
NICS	.36	.22	.13	.29	1.00

Table 95: Relative significance of ease of use items.

Criterium	E	R	G	DE	IC	ID	SP	SS	EU	NRS
E	1	2	5	6	6	7	7	5	3	.27
R	1/2	1	4	5	5	6	6	5	3	.22
G	1/5	1/4	1	2	2	3	3	2	1/3	.08
DE	1/6	1/5	1/2	1	1	2	2	1/2	1/4	.04
IC	1/6	1/5	1/2	1	1	3	3	1/2	1/3	.05
ID	1/7	1/6	1/3	1/2	1/3	1	1	1/3	1/5	.06
SP	1/7	1/6	1/3	1/2	1/3	1	1	1/3	1/5	.06
SS	1/5	1/5	1/2	2	2	3	3	1	1/3	.07
EU	1/3	1/3	3	4	3	5	5	3	1	.15
NICS	.36	.23	.07	.05	.05	.03	.03	.06	.12	1.00

Table 96: Relative significances of the performance criteria (efficiency most important).

Criterium	E	R	G	DE	IC	ID	SP	SS	EU	NRS
E	1	1	3	6	5	6	6	5	1/3	.20
R	1	1	3	6	5	6	6	5	1/3	.20
G	1/3	1/3	1	4	3	4	4	3	1/5	.12
DE	1/6	1/6	1/4	1	1/3	1	1	1/3	1/7	.02
IC	1/5	1/5	1/3	3	1	3	3	1	1/7	.07
ID	1/6	1/6	1/4	1	1/3	1	1	1/3	1/7	.02
SP	1/6	1/6	1/4	1	1/3	1	1	1/3	1/7	.02
SS	1/5	1/5	1/3	3	1	3	3	1	1/7	.07
EU	3	3	5	7	7	7	7	7	1	.28
NICS	.17	.17	.08	.03	.05	.03	.03	.05	.39	1.00

Table 97: Relative significances of the performance criteria (ease of use most important).

Criterium	E	R	G	DE	IC	ID	SP	SS	EU	NRS
E	1	1/4	1/2	5	5	5	4	4	3	.16
R	4	1	3	7	7	7	6	6	5	.28
G	2	1/3	1	6	6	6	5	5	4	.21
DE	1/5	1/7	1/6	1	1	1	1/3	1/3	1/4	.03
IC	1/5	1/7	1/6	1	1	1	1/3	1/3	1/4	.03
ID	1/5	1/7	1/6	1	1	1	1/3	1/3	1/4	.03
SP	1/4	1/6	1/5	3	3	3	1	1	1/3	.07
SS	1/4	1/6	1/5	3	3	3	1	1	1/3	.07
EU	1/3	1/5	1/4	4	4	4	3	3	1	.12
NICS	.13	.42	.19	.03	.03	.03	.05	.05	.07	1.00

Table 98: Relative significances of the performance criteria
(reliability most important).

References
•••

AB 1975 - AS 1973

AB 1975   J. Abadie, Methode du gradient reduit generalise: Le code
           GRGA, Note HI 1756/00, Électricité de France, Paris, 1975.

AB 1978   J. Abadie, The GRG method for non-linear programming,
           in: Design and implementation of optimization software,
           H.J. Greenberg ed., Sijthoff and Noordhoff, Alphen aan
           den Rijn, The Netherlands, 1978.

AC 1969   J. Abadie, J. Carpentier, The generalization of the
           Wolfe reduced gradient method to the case of nonlinear
           constraints, in: Optimization, R. Fletcher ed., Academic
           Press, 1969.

AG 1970   J. Abadie, J. Guigou, Numerical experiments with the GRG
           method, in: Integer and nonlinear programming, J. Abadie
           ed., North Holland Publishing Company, Amsterdam, 1970.

AH 1977   J. Abadie, A. Haggag, Methode quasi-Newtonienne dans une
           variante du gradient reduit generalise, Note HI 2458/00,
           Électricité de France, Paris, 1977.

AH 1979   J. Abadie, A. Haggag, Performance du gradient reduit
           generalise avec une methode quasi-Newtonienne pour la
           programmation non lineare, R.A.I.R.O. Recherche opératio-
           nelle / Operations Research, Vol.13, No.2 (1979),
           209-216.

AR 1966   L. Armijo, Minimization of functions having Lipschitz-
           continuous first partial derivatives, Pacific Journal on
           Mathematics, Vol.16 (1966), 1-3.

AS 1973   J. Asaadi, A computational comparison of some nonlinear
           programs, Mathematical Programming, Vol.4 (1973), 144-154.

BA 1975 – BI 1976

BA 1975   R. Bartels, Constrained least squares, quadratic programming, complementary pivot programming and duality,
Technical Report No.218, Department of Mathematical
Sciences, The Johns Hopkins University, Baltimore, USA,
1975.

BB 1978   M.J. Best, A.T. Bowler, ACDPAC: A FORTRAN IV subroutine
to solve differentiable mathematical programmes –
User's guide – Level 2.0, Research Report CORR 75-26,
Department of Combinatorics and Optimization, University
of Waterloo, Waterloo, Ontario, Canada, 1978.

BD 1970   Y. Bard, Comparison of gradient methods for the solution
of nonlinear parameter estimation, SIAM Journal on
Numerical Analysis, Vol.7, No.1 (1970), 157-187.

BE 1969   E. Beltrami, A comparison of some recent iterative
methods for the numerical solution of nonlinear programs, in: Computing methods in optimization problems,
Second International Conference, San Remo, 1968,
Springer-Verlag, Berlin, Heidelberg, New York, 1969.

BI 1971   M.C. Biggs, Computational experience with Murray's
algorithm for constrained minimization, Technical
Report No.23, Numerical Optimisation Centre, The
Hatfield Polytechnic, Hatfield, England, 1971.

BI 1972   M.C. Biggs, Constrained minimization using recursive
equality quadratic programming, in: Numerical methods
for non-linear optimization, F.A. Lootsma ed., Academic
Press, London, New York, 1971.

BI 1976   M.C. Bartholomew-Biggs, A numerical comparison between
two approaches to the nonlinear programming problem,
Technical Report No.77, Numerical Optimisation Centre,
The Hatfield Polytechnic, Hatfield, England, 1976.

BI 1978   M.C. Biggs, On the convergence of some constrained mini-
           mization algorithms based on recursive quadratic pro-
           gramming, Journal of the Institute of Mathematics and
           its Applications, Vol.21, No.1 (1978), 67-82.

BI 1979   M.C. Bartholomew-Biggs, An improved implementation of
           the recursive quadratic programming method for con-
           strained minimization, Technical Report No.105, Numeri-
           cal Optimisation Centre, The Hatfield Polytechnic,
           Hatfield, England, 1979.

BL 1967   E.M.L. Beale, An introduction to Beale's method of
           quadratic programming, in: Nonlinear programming,
           J. Abadie ed., North-Holland Publishing Company,
           Amsterdam, 1967.

BO 1966   M.J. Box, A comparison of several current optimization
           methods, and the use of transformations in constrained
           problems, Computer Journal, Vol.9 (1966), 67-77.

BR 1975   M.J. Best, K. Ritter, An accelerated conjugate direc-
           tion method to solve linearly constrained minimization
           problems, Journal of Computer and System Sciences,
           Vol.11, No.3 (1975), 295-322.

BR 1976   M.J. Best, K. Ritter, A class of accelerated conjugate
           direction methods for linearly constrained minimization
           problems, Mathematics of Computation, Vol.30, No.135
           (1976), 478-504.

BS 1975a   M.J. Best, FCDPAK: A FORTRAN IV subroutine to solve
           differentiable mathematical programmes - User's guide -
           Level 3.1, Research Report CORR 75-24, Department of
           Combinatorics and Optimization, University of Waterloo,
           Waterloo, Ontario, Canada, 1975.

BS 1975b   M.J. Best, A feasible conjugate direction method to solve linearly constrained optimization problems, Journal of Optimization Theory and Applications, Vol.16, No.1-2 (1975), 25-38.

BT 1973   R.P. Brent, Algorithms for minimizing without derivatives, Prentice-Hall, New Jersey, 1973.

BY 1967   C.G. Broyden, Quasi-Newton methods and their application to function minimization, Mathematics of Computation, Vol.21 (1967), 368-381.

BY 1970   C.G. Broyden, The convergence of a class of double-rank minimization algorithms, Journal of the Institute of Mathematics and its Applications, Vol.6 (1970), 76-90.

CG 1964   G.F. Coggins, Univariate search methods, Imperial Chemical Industries, Central Industrial Laboratory Research Note, 64/11, 1964.

CH 1978   L.W. Cornwell, P.A. Hutchison, M. Minkoff, H.K. Schultz, Test problems for constrained nonlinear mathematical programming algorithms, Technical Memorandum No.320, Argonne National Laboratory, Applied Mathematics Division, 1978.

CO 1968   A.R. Colville, A comparative study on nonlinear programming codes, IBM N.Y. Scientific Center Report 320-2949, 1968.

CM 1967   R.R. Coveyou, R.D. MacPherson, Fourier analysis of uniform random number, Journal of the Association of Computing Machinery, Vol.14 (1967), 100-119.

DA 1959 - DU 1974

DA 1959   W.C. Davidon, Variable metric method for minimization,
            A.E.C. Research and Development Report, ANL - 5990,
            1959.

DE 1976a   R.S. Dembo, A set of geometric programming test
            problems and their solutions, Mathematical Programming,
            Vol.10, No.2 (1976), 192-213.

DE 1976b   R.S. Dembo, The current state-of-the-art of algorithms
            and computer software for geometric programming,
            Working Paper 88, School of Organization and Manage-
            ment, Yale University, New Haven, 1976.

DI 1972   L.C.W. Dixon, The choice of steplength, a crutial
            factor on the performance of variable metric algorithms,
            in: Numerical methods for nonlinear optimization,
            F.A. Lootsma ed., Academic Press, London, New York,
            1972.

DL 1971   J.W. Daniel, The approximate minimization of functionals,
            Prentice Hall, Englewood Cliffs, New Jersey, 1971.

DM 1977   J.E. Dennis, J. Moré, Quasi-Newton methods, motivation
            and theory, SIAM Review, Vol.19 (1977), 46-89.

DN 1977   W.C. Davidon, L. Nazareth, DRVOCR - A FORTRAN implemen-
            tation of Davidon's optimally conditioned method,
            ANL - AMD Technical Memorandum No.306, Applied Mathe-
            matics Division, Argonne National Laboratory, 1977.

DP 1967   R.J. Duffin, E.L. Peterson, C. Zener, Geometric pro-
            gramming - Theory and application, John Wiley & Sons,
            New York, London, Sydney, 1967.

DU 1974   V. Dumitru, Gradient methods for unconstrained opti-
            mization, Econom. Comput. econom. Cybernetics Studies
            Res., No.4 (1974), 33-54.

DZ 1963   G. Dantzig, Linear programming and extensions,
             Princeton University Press, Princeton, 1963.

EF 1972   E.D. Eason, R.G. Fenton, Testing and evaluation of
             numerical methods for design optimization, Technical
             Publication Series UTME - TP 7204, Department of
             Mechanical Engineering, University of Toronto, 1972.

EI 1979   B. Einarsson, Bibliography on the evaluation of numeri-
             cal software, Journal of Computational and Applied
             Mathematics, Vol.5, No.2 (1979), 145-159.

FH 1965   P. Faure, P. Huard, Résolution des programmes mathéma-
             tiques à fonction nonlinéaire par la méthode du gradient
             réduit, Revue Francaise de Recherche Opérationalle,
             Vol.9 (1965), 167-205.

FL 1970   R. Fletcher, A new approach to variable metric algo-
             rithms, Computer Journal, Vol.13 (1970), 317-322.

FL 1971   R. Fletcher, A general quadratic programming algorithm,
             Journal of the Institute of Mathematics and its Appli-
             cations, Vol.7 (1971), 76-91.

FL 1975   R. Fletcher, An ideal penalty function for constrained
             optimization, in: Nonlinear programming 2, O.L. Manga-
             sarian, R.R. Meyer, S.M. Robinson eds., Academic Press,
             New York, 1975.

FM 1968   A.V. Fiacco, G.P. McCormick, Nonlinear sequential
             unconstrained minimization techniques, John Wiley &
             Sons, New York, 1968.

FP 1963   R. Fletcher, M.J.D. Powell, A rapidly convergent des-
             cent method for minimization, Computer Journal, Vol.6
             (1963), 163-168.

FR 1964   R. Fletcher, C.M. Reeves, Function minimization by con-
            jugate gradients, Computer Journal, Vol.7 (1964),
            149-154.

GA 1975   G.A. Gabriele, Application of the reduced gradient
            method to optimal engineering design, M.S. Thesis,
            School of Mechanical Engineering, Purdue University,
            1975.

GF 1969   D. Golfarb, Extension of Davidon's variable metric
            method to maximization under linear inequality and
            inequality constraints, SIAM Journal on Applied
            Mathematics, Vol.17 (1969), 739-764.

GF 1970   D. Goldfarb, A family of variable metric updates
            derived by variational means, Mathematics of Compu-
            tation, Vol.24 (1970), 23-26.

GI 1978   P.E. Gill, The design and implementation of software
            for unconstrained optimization, in: Design and imple-
            mentation of optimization software, H.J. Greenberg ed.,
            Sijthoff and Noordhoff, Alphen aan den Rijn, The
            Netherlands, 1978.

GM 1972   P.E. Gill, W. Murray, Quasi-Newton methods for uncon-
            strained optimization, Journal of the Institute of
            Mathematics and its Applications, Vol.9 (1972), 91-108.

GM 1974   P.E. Gill, W. Murray (eds.), Numerical methods for
            constrained optimization, Academic Press, New York,
            1974.

GM 1976   P.E. Gill, W. Murray, Minimization subject to bounds
            on the variables, National Physical Laboratory Report
            No. NAC 72, Teddington, Great Britain, 1976.

GM 1977   P.E. Gill, W. Murray, Numerically stable methods for quadratic programming, Report NAC 78, National Physical Laboratory, Teddington, Great Britain, 1977.

GO 1965   A. Goldstein, On steepest descent, SIAM Journal on Control, Vol.3 (1965), 147-151.

GR 1976   G.A. Gabriele, K.M. Ragsdell, OPT: A nonlinear programming code in FORTRAN-IV - User's manual, Modern Design Series, Vol.1, Purdue Research Foundation, Purdue University, West Lafayette, USA, 1976.

GS 1975   P.E. Gill, W. Murray, M.A. Saunders, Methods for computing and modifying the LDV factors of a matrix, Mathematics of Computation, Vol.29 (1975), 1051-1077.

HA 1975   H.Y. Huang, A.K. Aggarwal, A class of quadratically convergent algorithms for constrained function minimization, Journal of Optimization Theory and Applications, Vol.16, No.5/6 (1975), 447-486.

HB 1969   P.C. Haarhoff, J.D. Buys, H. von Molendorff, CONMIN: A computer programme for the minimization of a non-linear function subject to non-linear constraints, Report PEL 190, Atomic Energy Board, South Africa, 1969.

HB 1970   P.C. Haarhoff, J.D. Buys, A new method for the optimization of a nonlinear function subject to nonlinear constraints, Computer Journal, Vol.13 (1970), 178-184.

HE 1969   M.R. Hestenes, Multiplier and gradient methods, Journal of Optimization Theory and Applications, Vol.4 (1969), 303-320.

HI 1971 – KE 1973

HI 1971   D.M. Himmelblau, A uniform evaluation of unconstrained
           optimization techniques, in: Numerical methods for
           nonlinear optimization, F.A. Lootsma ed., Academic
           Press, London, New York, 1971.

HI 1972   D.M. Himmelblau, Applied nonlinear programming, McGraw-
           Hill Book – Company, New York, 1972.

HL 1970   H.Y. Huang, A.V. Levy, Numerical experiments on
           quadratically convergent algorithms for function
           minimization, Journal of Optimization Theory and
           Applications, Vol.6 (1970), 269-287.

HN 1976   S.-P. Han, Superlinearly convergent variable metric
           algorithms for general nonlinear programming problems,
           Mathematical Programming, Vol.11, No.3 (1976), 263-282.

HO 1969   A.G. Holzman, Comparative analysis of nonlinear pro-
           gramming codes with the Weisman algorithm, SRCC Report
           113, University of Pittsburgh, Pittsburgh, 1969.

HS 1980   W. Hock, K. Schittkowski, Nonlinear programming test
           examples, to appear: Lecture Notes in Economics and
           Mathematical Systems, Springer-Verlag.

IN 1972   J.P. Indusi, A computer algorithm for constrained
           minimization, in: Minimization algorithms, G.P. Szegö
           ed., Academic Press, New York, London, 1972.

KB 1968   J. Kowalik, M.R. Osborne, Methods for unconstrained
           optimization problems, American Elsevier Publishing
           Co., New York, 1968.

KE 1973   E.L. Keller, The general quadratic optimization problem,
           Mathematical Programming, Vol.5, No.3 (1973), 311-337.

KM 1973   J.L. Kuester, J.H. Mize, Optimization techniques with
           FORTRAN, McGraw-Hill Book - Company, New York, 1973.

KO 1977   D. Koo, Elements of optimization, Springer-Verlag,
           New York, Heidelberg, Berlin, 1977.

KR 1977   D. Kraft, Nichtlineare Programmierung - Grundlagen,
           Verfahren, Beispiele, Forschungsbericht, DFVLR, Ober-
           pfaffenhofen, Germany, F.R., 1977.

LH 1974   C.L. Lawson, R.J. Hanson, Solving least squares problems,
           Prentice Hall, Englewood Cliffs, New Jersey, 1974.

LJ 1978   L.S. Lasdon, A.D. Waren, A. Jain, M. Ratner, Design and
           testing of a generalized reduced gradient code for
           nonlinear programming, ACM Transactions on Mathemati-
           cal Software, Vol.4, No.1 (1978), 34-50.

LO 1971   F.A. Lootsma, A survey of methods for solving con-
           strained minimization problems via unconstrained
           minimization, in: Numerical methods for nonlinear
           optimization, F.A. Lootsma ed., Academic Press, London,
           New York, 1971.

LO 1974   F.A. Lootsma, Convergence rates of quadratic exterior
           penalty-function methods for solving constrained
           minimization problems, Philips Research Report No.29,
           1974.

LO 1978   F.A. Lootsma, The ALGOL 60 procedure MINIFUN for sol-
           ving nonlinear optimization problems, in: Design and
           implementation of optimization software, H.J. Greenberg
           ed., Sijthoff and Noordhoff, Alphen aan den Rijn, The
           Netherlands, 1978.

LO 1979   F.A. Lootsma, Performance evaluation of non-linear
           programming codes from the viewpoint of a decision
           maker, in: Performance evaluation of numerical soft-
           ware, L.D. Fosdick ed., North Holland Publishing
           Company, Amsterdam, New York, Oxford, 1979.

LP 1973   A. Land, S. Powell, FORTRAN codes for mathematical
           programming, John Wiley & Sons, Chichester, New York,
           Brisbane, Toronto, 1973.

LR 1978   L.S. Lasdon, A.D. Waren, M.W. Ratner, GRG2 user's
           guide, Report, School of Business Administration, The
           University of Texas at Austin, Austin, Texas, 1978.

LU 1965   D.G. Luenberger, Introduction to linear and nonlinear
           programming, Addison-Wesley Publishing Company, Reading,
           Massachusetts, 1965.

LW 1978   L.S. Lasdon, A.D. Waren, Generalized reduced gradient
           software for linearly and nonlinearly constrained
           problems, in: Design and implementation of optimiza-
           tion software, H.J. Greenberg ed., Sijthoff and
           Noordhoff, Alphen aan den Rijn, The Netherlands, 1978.

MC 1967   G.P. McCormick, Second order conditions for constrained
           minima, SIAM Journal on Applied Mathematics, Vol.15,
           No.3 (1967), 641-652.

MD 1976   B. Murtagh, M. Saunders, Nonlinear programming for
           large sparse systems, Technical Report SOL 76-15,
           Department of Operations Research, Stanford University,
           Stanford, California, 1976.

MG 1978   J.J. Moré, B.S. Garbow, K.E. Hillstrom, Testing uncon-
           strained optimization software, Technical Memorandum
           No.324, Argonne National Laboratory, Applied Mathe-
           matics Division, Argonne, 1978.

MH 1971   W.C. Mylander, R.L. Holmes, G.P. McCormick, A guide
           to SUMT-Version 4: The computer program implementing
           the sequential unconstrained minimization technique
           for nonlinear programming, Research Analysis Corpo-
           ration, Mc Lean, Virginia, 1971.

MK 1975   J.J. McKeown, Specialized versus general-purpose
           algorithms for minimizing functions that are sums of
           squared terms, Mathematical Programming, Vol.9 (1975),
           57-68.

MS 1969   B.A. Murtagh, R.W.H. Sargent, A constrained minimization
           method with quadratic convergence, in: Optimization,
           R. Fletcher ed., Academic Press, London, New York, 1969.

MT 1972   A. Miele, J.L. Tietze, A.V. Levy, Comparison of several
           gradient algorithms for mathematical programming prob-
           lems, Aero-Astronautics Report No.94, Rice University,
           Houston, Texas, 1972.

MU 1967   W. Murray, Ill-conditioning in barrier and penalty
           functions arising in constrained nonlinear programming,
           in: Proceedings of the Sixth International Symposium
           on Mathematical Programming, Princeton University,
           Princeton, New Jersey, 1967.

MU 1969a   W. Murray, Constrained optimization, National Physical
           Laboratory Report MA 79, Teddington, Great Britain,
           1969.

MU 1969b   W. Murray, An algorithm for constrained minimization,
           in: Optimization, R. Fletcher ed., Academic Press,
           London, New York, 1969.

MU 1976   W. Murray, Methods for constrained optimization, in:
           Optimization in action, L.C.W. Dixon ed., Academic
           Press, New York, 1976.

NA 1978   L. Nazareth, Software for optimization, Technical
Report SOL 78-32, Systems Optimization Laboratory,
Stanford University, Stanford, California, 1978.

NH 1975   J.S. Newell, D.M. Himmelblau, A new method for nonlinearly constrained optimization, AIChE Journal,
Vol.21, No.3 (1975), 479-486.

OR 1970   J.M. Ortega, W.C. Rheinboldt, Iterative solution of
nonlinear equations in several variables, Academic
Press, New York, San Francisco, London, 1970.

PI 1969   D.A. Pierre, Optimization theory with applications,
John Wiley & Sons, New York, London, Sydney, Toronto,
1969.

PL 1975   D.A. Pierre, M.J. Lowe, Mathematical programming via
augmented Lagrangians. An introduction with computer
programs, Addison-Wesley Publishing Company, Reading,
Massachusetts, 1975.

PO 1964   M.J.D. Powell, An efficient method for finding the
minimum of a function of deveral variables without
calculating derivatives, The Computer Journal, Vol.7
(1964), 155-162.

PO 1969   M.J.D. Powell, A method for nonlinear constraints in
minimization problems, in: Optimization, R. Fletcher
ed., Academic Press, London, New York, 1969.

PO 1974   M.J.D. Powell, Introduction to constrained optimization,
in: Numerical methods for constrained optimization,
P.E. Gill, W. Murray eds., Academic Press, London,
New York, San Francisco, 1974.

PO 1978a   M.J.D. Powell, A fast algorithm for nonlinearly constrained optimization calculations, in: Proceedings
of the 1977 Dundee Conference on Numerical Analysis,
Lecture Notes in Mathematics, Springer-Verlag, Berlin,
Heidelberg, New York, 1978.

PO 1978b   M.J.D. Powell, The convergence of variable metric
methods for nonlinearly constrained optimization
calculations, in: Nonlinear programming 3, O.L. Manga-
sarian, R.R. Meyer, S.M. Robinson eds., Academic Press,
New York, San Francisco, London, 1978.

PP 1975   J. Peschon, N.M. Peterson, Optimization and simulation
computations in advanced energy control centers for
electric power utilities, in: Proceedings of the XX
International Meeting, The Institute of Management
Sciences, on Management Sciences Developing Countries,
and National Priorities, Jerusalem Academic Press, 1975.

RB 1972   S.M. Robinson, A quadratically convergent algorithm
for general nonlinear programming problems, Mathematical Programming, Vol.3 (1972), 145-155.

RI 1977   M.J. Rijckaert, Computational aspects of geometric
programming, in: Design and implementation of optimization software, H.J. Greenberg ed., Sijthoff and
Noordhoff, Alphen aan den Rijn, The Netherlands, 1978.

RM 1978   M.J. Rijckaert, X.M. Martens, A comparison of generalized geometric programming algorithms, Journal of
Optimization Theory and Applications, Vol.26, No.2
(1978), 205-242.

RO 1974   R.T. Rockafellar, Augmented Lagrange multiplier functions and duality in non-convex programming, SIAM
Journal on Control, Vol.12 (1974), 268-285.

RR 1978   R.R. Root, K.M. Ragsdell, BIAS: A nonlinear programming
           code in FORTRAN-IV: User's manual, Report, School of
           Mechanical Engineering, Purdue University, West
           Lafayette, Indiana, 1978.

RT 1977   R.R. Root, An investigation of the method of multipliers
           for engineering design, Ph.D. Dissertation, Purdue
           University, West Lafayette, Indiana, 1977.

RU 1978   D. Rufer, User's guide for NLP - A subroutine package
           to solve nonlinear optimization problems, Report No.
           78-07, Fachgruppe für Automatik, Eidgenössische
           Technische Hochschule, Zürich, Switzerland, 1978.

RU 1979   D. Rufer, Implementation and properties of a method
           for the identification of nonlinear continuous time
           models, Proceedings of the 7th IFAC World Congress
           1978, A. Niemi ed., Pergamon Press, 1979.

RW 1977   H. Ramsin, P.A. Wedin, A comparison of some algorithms
           for the nonlinear least squares problem, BIT, Vol.17,
           No.1 (1977), 72-90.

SA 1977   E. Sandgren, The utility of nonlinear programming
           algorithms, Ph.D. Thesis, Purdue University, West
           Lafayette, Indiana, 1977.

SC 1979   K. Schittkowski, The construction of degenerate, ill-
           conditioned, and indefinite nonlinear programming
           problems and their usage to test optimization programs,
           submitted for publication.

SG 1977   S.B. Schuldt, G.A. Gabriele, R.R. Root, E. Sandgren,
           K.M. Ragsdell, Application of a new penalty function
           method to design optimization, Journal of Engineering
           for Industry, Trans. ASME, Series B, Vol.99, No.1
           (1977), 31-36.

SH 1973   R.L. Staha, Constrained optimization via moving ex-
           terior truncations, Ph.D. Thesis, The University of
           Texas, Austin, Texas, 1973.

SO 1970   D.F. Shanno, Conditioning of quasi-Newton methods for
           function minimization, Mathematics of Computation,
           Vol.24 (1970), 647-656.

SS 1979   K. Schittkowski, J. Stoer, A factorization method for
           the solution of constrained linear least squares
           problems allowing subsequent data changes, Numerische
           Mathematik, Vol.31, Fasc.4 (1979), 431-463.

ST 1969   D.C. Stocker, A comparative study of nonlinear pro-
           gramming codes, M.S. Thesis, The University of Texas,
           Austin, Texas, 1969.

SY 1975   T.L. Saaty, Hierarchies and priorities - eigenvalue
           analysis, Internal Report, University of Pennsylvania,
           Wharton School, Philadelphia, 1975.

SY 1977   T.L. Saaty, A scaling method for priorities in hierar-
           chical structures, Journal on Mathematical Psycholo-
           gy, Vol.15 (1977), 234-281.

SZ 1977   K. Schittkowski, P. Zimmermann, A factorization method
           for constrained least squares problems with data
           changes. Part 2: Numerical tests, comparisons, and
           ALGOL codes, Preprint No.30, Institut für Angewandte
           Mathematik und Statistik, Universität Würzburg, Würzburg,
           Germany, F.R., 1977.

TA 1969   D. Tabak, Comparative study of various minimization
           techniques used in mathematical programming, IEEE
           Transactions on Automatic Control, AC-14 (1969), 572.

WI 1963   R.B. Wilson, A simplicial algorithm for concave pro-
           gramming, Ph.D. Dissertation, Graduate School of
           Business Administration, Harward University, Boston,
           1963.

WL 1979   A.D. Waren, L.S. Lasdon, The status of nonlinear
           programming software, Operations Research, Vol.27,
           No.3 (1979), 431-456.

WO 1959   P. Wolfe, The simplex method for quadratic programming,
           Econometrica, Vol.27 (1959), 382-398.

WO 1967   P. Wolfe, Methods of nonlinear programming, in: Recent
           advances in mathematical programming, R.L. Graves,
           P. Wolfe eds., McGraw-Hill Book-Company, New York, 1963.

WO 1967   P. Wolfe, Methods for linear constraints, in: Nonlinear
           programming, J. Abadie ed., North-Holland Publishing
           Company, Amsterdam, 1967.

ZI 1974   G. Zielke, Test matrices with maximal condition number,
           Computing, Vol.13 (1974), 33-54.

Class	1A	2A	3A	4A	5A	6A	7A	8A	9A	10A
FV	.86E-8	.61E-7	.16E-7	.25E-8	.24E-8	.28E-7	.17E-7	.20E-7	.12E-7	.62E-6
VC	.15E-7	.91E-7	.21E-7	.34E-9	.36E-8	.50E-8	.23E-7	.31E-7	.86E-9	.70E-6
KT	.26E-8	.18E-8	.91E-9	.23E-5	.28E-5	.25E-6	.13E-6	.13E-7	.90E-5	.12E-4
ED	9.37	10.43	9.66	6.70	6.61	7.30	7.19	9.08	4.90	6.15
ET	10.2	23.3	16.6	12.7	13.3	4.1	38.2	18.0	96.6	66.4
NF	90	77	28	37	33	50	30	26	131	172
NG	541	931	86	372	336	406	390	377	658	3099
NDF	53	50	28	30	29	36	28	26	71	99
NDG	321	604	84	300	296	293	364	377	360	1799
PNS	60.0	66.7	25.0	0	0	0	26.7	0	20.0	20.0
FFV	.22E-10	.72E-5	.61E-9	-	-	-	.11	-	.0	.20E-5
FVC	.47E+1	.42E+2	.20E-1	-	-	-	.53E-5	-	.43E+4	.15E-2
F	0	0	1	0	0	0	0	0	0	0
PGS	50.0	0	80.0	0	0	0	90.9	0	0	25.0
GFV	-.19	-	-.33E+1	-	-	-	-.18	-	-	-.15E-2
GVC	.44E-7	-	.20E-7	-	-	-	.18E-7	-	-	.49E-8

Table 33: Test results for OPRQP (general problems).

Class	1B	2B	3B	4B	5B	6B	7B	8B	9B	10B
FV	.17E-9	.11E-9	.21E-10	.10E-11	.65E-9	.58E-10	.10E-10	.47E-8	.15E-8	.39E-9
VC	.16E-9	.12E-9	.18E-10	.73E-11	.12E-8	.81E-10	.16E-11	.73E-8	.22E-8	.60E-8
KT	.15E-6	.19E-7	.57E-8	.87E-8	.87E-6	.74E-7	.16E-6	.28E-8	.22E-8	.11E-8
ED	7.77	8.54	9.02	8.85	7.29	8.08	6.26	9.67	9.66	9.38
A	8.54	9.03	9.67	10.01	7.86	8.89	8.96	8.67	8.95	8.74
ET	13.4	13.3	14.8	12.5	11.5	13.7	26.9	12.0	12.7	11.9
NF	31	32	35	36	26	32	82	25	27	25
NG	310	321	357	368	265	327	825	257	275	258
NDF	28	27	31	25	24	29	53	25	27	25
NDG	281	278	311	255	245	290	538	255	270	257
ET/A	1.6	1.5	1.5	1.3	1.5	1.6	3.0	1.4	1.4	1.4
NF/A	3.6	3.6	3.7	3.7	3.4	3.7	9.1	3.0	3.1	3.0
NG/A	36.1	35.7	36.7	37.4	33.5	36.9	91.3	29.6	30.7	29.5
NDF/A	3.3	3.1	3.2	2.6	3.1	3.2	6.0	3.0	3.0	2.9
NDG/A	32.8	30.9	32.1	25.6	31.1	32.7	59.9	29.4	30.1	29.4
R	0	0	0	0	0	0	0	0	1	0

<u>Table 34:</u> Test results for OPRQP (degenerate, ill-conditioned, and indefinite problems).

Class	1A	2A	3A	4A	5A	6A	7A	8A	9A	10A
FV	.14E-6	.10E-8	.68E-8	.87E-7	.18E-7	.70E-7	.40E-7	.13E-7	.28E-4	.13E-6
VC	.50E-7	.26E-7	.93E-8	.63E-8	.79E-8	.14E-7	.53E-7	.20E-7	.12E-6	.22E-6
KT	.39E-6	.23E-8	.48E-8	.13E-3	.48E-4	.34E-4	.63E-6	.91E-8	.30E-2	.21E-6
ED	7.96	10.89	9.35	5.21	5.48	5.56	6.83	9.20	2.81	8.03
ET	20.2	24.7	13.3	11.8	12.2	5.5	33.1	15.2	35.1	19.6
NF	192	87	20	25	24	64	23	19	29	30
NG	1155	1044	60	258	244	514	299	273	147	546
NDF	81	51	20	24	23	38	21	19	24	27
NDG	486	612	60	249	238	309	273	273	120	486
PNS	56.7	97.7	43.3	6.7	0	0	20.0	0	26.7	46.7
FFV	.10E-8	.13E-4	.36E-9	.14E-2	-	-	.78E-1	-	.13E-5	.54E-9
FVC	.13E+1	.28E+4	.16E+2	.25E-7	-	-	.31E-4	-	.54E+2	.39E-1
F	0	0	0	0	0	0	0	0	0	0
PGS	53.8	0	64.7	0	0	0	91.7	0	0	0
GFV	-.22	-	-.10E+1	-	-	-	-.19	-	-	-
GVC	.46E-8	-	.76E-8	-	-	-	.83E-8	-	-	-

Table 35: Test results for XROP (general problems).

Class	1B	2B	3B	4B	5B	6B	7B	8B	9B	10B
FV	.25E-7	.36E-7	.79E-7	.32E-6	.49E-8	.21E-6	.61E-6	.35E-7	.88E-7	.48E-7
VC	.14E-7	.27E-8	.31E-8	.24E-9	.85E-8	.30E-8	.75E-8	.79E-8	.53E-8	.79E-8
KT	.16E-3	.19E-3	.27E-3	.47E-3	.15E-4	.25E-3	.41E-3	.33E-4	.18E-3	.82E-4
ED	4.99	4.38	4.31	3.86	6.04	4.18	3.19	6.29	6.16	6.26
A	6.06	6.03	5.88	5.83	6.81	5.75	5.23	6.58	6.31	6.44
ET	7.7	8.2	7.8	6.7	7.4	8.5	8.7	8.1	8.2	8.3
NF	15	16	16	14	14	16	17	15	15	15
NG	155	165	160	142	141	165	171	156	157	156
NDF	14	15	15	12	14	16	16	15	15	15
NDG	146	156	150	127	141	162	166	155	156	155
ET/A	1.3	1.4	1.3	1.2	1.1	1.5	1.7	1.2	1.3	1.3
NF/A	2.6	2.7	2.7	2.5	2.1	2.8	3.4	2.4	2.5	2.4
NG/A	25.7	27.3	27.4	24.8	20.8	28.4	33.5	23.9	25.0	24.3
NDF/A	2.4	2.6	2.6	2.2	2.1	2.8	3.2	2.4	2.5	2.4
NDG/A	24.3	25.9	25.7	22.1	20.8	27.9	32.4	23.7	24.8	24.2
R	0	0	0	0	0	0	0	0	0	0

Table 36: Test results for XROP (degenerate, ill-conditioned, and indefinite problems).

Class	1A	2A	3A	4A	5A	6A	7A	8A	9A	10A
FV	.69E-7	.17E-11	.92E-8	.44E-7	.60E-8	.23E-7	.70E-8	.80E-8	—	.17E-8
VC	.26E-7	.18E-11	.21E-7	.11E-11	.10E-11	.76E-9	.12E-7	.18E-9	—	.13E-7
KT	.51E-5	.16E-11	.12E-4	.14E-3	.11E-3	.16E-3	.32E-4	.68E-5	—	.40E-5
ED	6.69	11.96	5.97	5.15	5.30	5.28	5.47	6.54	—	6.38
ET	4.7	5.3	11.2	23.3	31.0	23.2	55.0	53.7	—	153.5
NF	16	8	12	18	19	19	14	22	—	20
NG	96	96	37	184	191	157	192	320	—	362
NDF	16	8	12	18	19	19	14	22	—	20
NDG	96	96	37	184	191	157	192	320	—	362
PNS	6.7	10.0	23.3	3.3	0	0	0	0	0	8.3
FFV	.59E+1	.11E+1	.53E-2	.18E-2	—	—	—	—	—	.37E-2
FVC	.19E-3	.47E-6	.37E-1	.0	—	—	—	—	—	.37E-8
F	0	0	0	0	0	0	0	0	4	1
PGS	53.6	7.4	87.0	0	0	6.7	66.7	0	100.0	18.2
GFV	-.40E+1	-.20E+1	-.20E+1	—	—	-.17E-1	-.13	—	-.34E+2	-.16E-2
GVC	.91E-9	.37E-6	.14E-8	—	—	.56E-8	.57E-9	—	.31E-4	.75E-9

Table 37: Test results for VF02AD (general problems).

Class	1B	2B	3B	4B	5B	6B	7B	8B	9B	10B
FV	.25E-7	.68E-7	.93E-7	.12E-6	.47E-7	.78E-7	.42E-6	.15E-6	.16E-6	.16E-6
VC	.14E-11	.0	.0	.0	.0	.12E-11	.0	.0	.0	.0
KT	.20E-3	.26E-3	.32E-3	.30E-3	.25E-3	.17E-3	.25E-3	.17E-3	.16E-3	.19E-3
ED	4.77	4.27	4.23	3.96	4.82	4.35	3.19	6.48	6.45	6.45
A	6.98	6.76	6.69	6.60	6.94	6.78	6.29	7.27	7.26	7.25
ET	23.0	22.5	21.1	24.0	14.3	22.8	22.8	13.2	12.5	12.7
NF	13	13	12	14	8	12	13	10	9	10
NG	131	131	123	141	80	127	130	100	98	100
NDF	13	13	12	14	8	12	13	10	9	10
NDG	131	131	123	141	80	127	130	100	98	100
ET/A	3.3	3.3	3.1	3.6	2.0	3.3	3.6	1.8	1.7	1.7
NF/A	1.9	2.0	1.9	2.1	1.1	1.9	2.1	1.3	1.3	1.4
NG/A	18.8	19.4	18.4	21.3	11.4	18.5	20.7	13.5	13.4	13.6
NDF/A	1.9	2.0	1.9	2.1	1.1	1.9	2.1	1.3	1.3	1.4
NDG/A	18.8	19.4	18.4	21.3	11.4	18.5	20.7	13.5	13.4	13.6
R	0	0	0	0	0	0	0	0	0	0

Table 38: Test results for VF02AD (degenerate, ill-conditioned, and indefinite problems).

Class	1A	2A	3A	4A	5A	6A	7A	8A	9A	10A
FV	.83E-7	.10E-7	.52E-6	.38E-4	.75E-4	.26E-4	.33E-10	.21E-3	.50E-4	.13E-4
VC	.25E-9	.97E-11	.25E-9	.11E-11	.10E-11	.15E-11	.21E-10	.14E-11	.24E-11	.59E-11
KT	.28E-3	.37E-8	.97E-3	.41E-2	.65E-2	.71E-2	.40E-9	.38E-1	.59E-2	.99E-2
ED	5.25	9.93	4.11	3.35	2.90	3.86	11.53	2.77	2.81	4.08
ET	16.6	51.0	25.5	52.2	33.4	6.3	7.2	41.7	89.2	47.7
NF	137	171	99	286	258	146	10	165	163	201
NG	1851	3985	457	4363	2077	1526	273	3236	1585	5670
NDF	61	72	34	78	76	50	2	67	79	70
NDG	168	367	57	456	470	232	39	432	260	641
PNS	3.3	24.7	3.3	3.3	3.3	20.0	24.7	0	0	40.0
FFV	.33E+1	.62	.65E-3	.17E-2	.80E-3	.11E+1	.13E-2	–	–	.61E-10
FVC	.27E-11	.0	.23E-9	.13E-11	.0	.17E-4	.14E-11	–	–	.26E+1
F	0	0	0	0	0	0	0	0	3	0
PGS	65.5	13.6	86.2	0	0	4.2	90.9	0	0	0
GFV	-.69E+1	-.20E+1	-.14E+3	–	–	-.17E-1	-.11	–	–	–
GVC	.21E-9	.0	.15E-9	–	–	.0	.31E-11	–	–	–

Table 39: Test results for GRGA (general problems).

Class	1B	2B	3B	4B	5B	6B	7B	8B	9B	10B
FV	.12E-3	.15E-3	.10E-3	.42E-4	.22E-3	.42E-3	.36E-3	.87E-6	.83E-6	.88E-6
VC	.11E-11	.14E-11	.0	.0	.0	.11E-10	.20E-11	.31E-10	.23E-10	.26E-10
KT	.13E-1	.19E-1	.15E-1	.73E-2	.17E-1	.23E-1	.18E-1	.30E-3	.31E-3	.34E-3
ED	2.97	2.83	2.98	2.88	3.04	2.64	2.38	5.74	5.76	5.76
A	5.18	5.05	5.20	5.35	5.11	4.91	4.82	6.46	6.50	6.47
ET	18.2	23.1	21.5	39.1	10.4	8.3	6.7	12.0	11.9	11.9
NF	109	137	137	184	57	42	32	62	62	62
NG	1135	1786	1326	3816	551	448	385	662	663	663
NDF	40	41	45	54	23	18	12	32	32	32
NDG	228	240	273	292	145	117	91	160	158	158
ET/A	3.5	4.7	4.1	7.6	2.0	1.7	1.4	1.9	1.8	1.8
NF/A	21.0	27.5	26.6	35.5	11.1	8.6	6.8	9.6	9.6	9.6
NG/A	220.6	364.0	255.5	748.1	107.3	91.4	80.8	102.2	101.7	102.2
NDF/A	7.7	8.4	8.6	10.5	4.5	3.7	2.7	5.0	4.9	4.9
NDG/A	44.3	47.7	52.6	55.7	28.2	23.8	19.0	24.7	24.3	24.4
R	0	0	0	0	0	1	0	0	0	0

Table 40: Test results for GRGA (degenerate, ill-conditioned, and indefinite problems).

Class	1A	2A	3A	4A	5A	6A	7A	8A	9A	10A
FV	.90E-7	.25E-7	.42E-4	.24E-3	.21E-3	.17E-4	-	.16E-4	-	.14E-4
VC	.15E-7	.55E-7	.55E-6	.33E-8	.11E-9	.39E-10	-	.62E-7	-	.20E-7
KT	.10E-3	.59E-8	.82E-2	.59E-2	.12E-1	.42E-2	-	.11E-1	-	.10E-1
ED	5.69	10.20	2.94	2.87	2.53	3.75	-	3.21	-	3.85
ET	16.5	52.4	60.1	58.4	56.9	19.0	-	99.3	-	168.2
NF	155	116	420	655	717	760	-	831	-	1372
NG	2325	5370	1483	5231	5531	5381	-	11104	-	21190
NDF	0	0	0	0	0	0	-	0	-	0
NDG	45	99	63	350	381	273	-	373	-	567
PNS	50.0	70.8	23.7	66.7	56.7	16.7	51.7	13.7	100.0	46.7
FFV	.50E-2	.51E+3	.30E-10	.29E-1	.11E-1	.18	.17E-2	.11E-1	.11E+1	.65
FVC	.14	.62E-9	.21	.88E-10	.13E-10	.21E-7	.68E-2	.25E-9	.87E-4	.22E-9
F	2	2	0	0	0	0	1	0	2	0
PGS	83.3	42.9	78.3	0	0	4.0	100.0	0	0	0
GFV	-.32E+2	-.20E+1	-.43E+1	-	-	-.17E-1	-.16	-	-	-
GVC	.12E-7	.16E-9	.67E-6	-	-	.58E-10	.18E-6	-	-	-

Table 41: Test results for OPT (general problems).

Class	1B	2B	3B	4B	5B	6B	7B	8B	9B	10B
FV	.28E-3	.40E-3	.25E-3	.26E-3	.53E-4	.38E-4	.13E-3	.29E-5	.52E-5	.29E-5
VC	.26E-6	.10E-8	.13E-8	.25E-9	.51E-9	.90E-9	.44E-8	.17E-7	.23E-8	.58E-8
KT	.18E-1	.26E-1	.21E-1	.22E-1	.77E-2	.62E-2	.11E-1	.62E-3	.98E-3	.63E-3
ED	2.77	2.66	2.68	2.71	3.12	3.22	2.63	5.40	5.26	5.38
A	3.66	4.16	4.21	4.38	4.70	4.72	4.21	5.48	5.55	5.59
ET	25.4	38.2	30.7	52.7	38.2	37.2	16.6	25.3	24.7	21.4
NF	310	484	373	675	483	478	182	292	288	246
NG	2388	3378	2936	4996	3865	3218	1732	2276	2218	1927
NDF	0	0	0	0	0	0	0	0	0	0
NDG	148	248	171	298	185	254	64	150	153	138
ET/A	7.1	8.1	7.4	11.9	7.9	8.0	3.7	4.6	4.3	3.7
NF/A	86.4	100.7	89.7	152.6	100.1	102.6	40.2	53.1	50.2	42.6
NG/A	676.9	728.7	725.8	1139.4	805.3	696.9	386.9	414.5	388.9	336.1
NDF/A	.0	.0	.0	.0	.0	.0	.0	.0	.0	.0
NDG/A	40.4	51.3	38.5	66.5	38.5	53.7	14.4	27.3	26.9	24.0
R	2	2	2	2	0	0	1	0	0	0

Table 42: Test results for OPT (degenerate, ill-conditioned, and indefinite problems).

Class	1A	2A	3A	4A	5A	6A	7A	8A	9A	10A
FV	.52E-4	.48E-6	.59E-6	.17E-7	.81E-8	.32E-5	.26E-4	.47E-4	.20E-6	.16E-4
VG	.29E-4	.12E-6	.33E-6	.75E-9	.92E-9	.44E-6	.12E-3	.34E-7	.13E-6	.49E-4
KT	.25E-2	.70E-7	.11E-3	.25E-4	.16E-4	.17E-3	.17E-2	.98E-3	.83E-4	.48E-3
ED	4.32	8.86	5.03	5.91	5.98	5.07	3.59	4.19	4.90	5.06
ET	12.0	24.8	47.9	33.7	33.8	6.5	146.3	45.7	216.5	172.7
NF	183	141	314	229	251	153	445	209	693	948
NG	1100	1703	2175	2295	2519	1226	5785	2938	3465	17076
NDF	15	13	57	28	32	27	50	37	109	104
NDG	93	165	245	286	329	218	650	519	550	1889
PNS	13.3	13.3	16.7	6.3	0	0	33.3	0	26.3	6.7
FFV	.28E-1	.17E+3	.45E+6	.17E-2	-	-	.34E-2	-	.83E-9	.37E-2
FVC	.19E-3	.10E-4	.36E-4	.12E-8	-	-	.14E-4	-	.22E-1	.22E-5
F	0	0	0	0	0	0	0	0	0	0
PGS	65.4	3.8	68.0	0	0	.0	80.0	0	0	0
GFV	-.15E+1	-.20E+1	-.14E+2	-	-	-	-.23	-	-	-
GVC	.53E-5	.55E-4	.28E-4	-	-	-	.42E-4	-	-	-

Table 43: Test results for GRG2(1) (general problems).

Class	1B	2B	3B	4B	5B	6B	7B	8B	9B	10B
FV	.73E-8	.39E-8	.28E-7	.21E-8	.28E-8	.26E-6	.89E-6	.35E-5	.61E-5	.37E-5
VC	.32E-10	.54E-10	.11E-8	.41E-10	.96E-11	.20E-8	.37E-10	.98E-6	.11E-6	.11E-5
KT	.95E-5	.77E-4	.66E-4	.48E-4	.46E-5	.27E-3	.99E-4	.31E-4	.78E-4	.18E-4
ED	6.11	5.38	5.18	5.11	6.68	4.16	2.76	5.42	5.31	5.42
A	7.44	7.04	6.46	7.13	7.89	5.75	5.81	5.35	5.40	5.39
ET	17.3	17.1	15.2	19.4	16.6	18.2	22.2	8.5	8.5	8.5
NF	122	119	103	143	122	132	179	37	38	38
NG	1220	1198	1036	1438	1226	1323	1793	378	380	382
NDF	15	15	14	16	14	15	16	10	10	10
NDG	157	157	146	163	141	157	160	106	106	106
ET/A	2.3	2.5	2.4	2.7	2.1	3.9	4.0	1.6	1.6	1.6
NF/A	16.1	17.3	16.2	19.7	15.2	29.0	32.0	7.1	7.3	7.1
NG/A	160.7	172.5	162.0	197.2	152.3	289.9	319.9	71.1	73.2	71.0
NDF/A	2.1	2.3	2.3	2.3	1.8	3.2	2.9	2.0	2.0	2.0
NDG/A	21.0	22.7	23.0	22.7	17.9	31.7	28.8	19.9	20.4	19.7
R	0	0	0	0	0	0	0	0	0	0

Table 44: Test results for GRG2(1) (degenerate, ill-conditioned, and indefinite problems).

Class	1A	2A	3A	4A	5A	6A	7A	8A	9A	10A
FV	.12E-3	.29E-4	.27E-4	.69E-5	.68E-7	.15E-5	.82E-6	.25E-6	.17E-4	.91E-5
VC	.13E-3	.68E-4	.52E-4	.28E-9	.87E-10	.10E-6	.16E-5	.12E-7	.16E-6	.24E-4
KT	.70E-3	.68E-5	.40E-3	.31E-2	.40E-3	.83E-3	.13E-3	.23E-3	.40E-2	.12E-2
ED	4.87	7.23	4.23	3.97	4.71	4.59	4.70	4.87	3.13	4.97
ET	13.8	18.7	43.8	49.6	54.9	13.0	238.6	134.6	216.9	350.7
NF	245	128	354	471	670	466	1300	1104	1888	2565
NG	1473	1542	1063	4715	6707	3734	16900	15461	9444	46183
NDF	0	0	0	0	0	0	0	0	0	0
NDG	0	0	0	0	0	0	0	0	0	0
PNS	56.7	60.0	13.3	6.7	0	0	8.3	0	26.7	33.3
FFV	.76E-10	.14E-3	.33E+3	.17E-2	-	-	.20E-1	-	.84E-9	.30E-2
FVC	.38E-2	.84E-2	.25E-4	.12E-8	-	-	.10E-5	-	.44E-2	.80E-5
F	0	5	0	0	0	0	1	0	0	3
PGS	84.6	0	88.5	0	0	0	90.9	0	0	0
GFV	-.40	-	-.26E+2	-	-	-	-.12	-	-	-
GVC	.11E-4	-	.14E-4	-	-	-	.67E-6	-	-	-

Table 45: Test results for GRG2(2) (general problems).

Class	1B	2B	3B	4B	5B	6B	7B	8B	9B	10B
FV	.61E-7	.29E-6	.30E-6	.39E-6	.34E-7	.18E-6	.19E-5	.34E-5	.59E-5	.37E-5
VC	.13E-9	.20E-9	.74E-9	.21E-10	.25E-10	.31E-9	.47E-9	.99E-6	.11E-6	.11E-5
KT	.54E-3	.88E-3	.90E-3	.11E-2	.20E-3	.28E-3	.35E-3	.36E-3	52E-3	.34E-3
ED	4.66	4.24	4.16	4.08	4.92	4.26	2.69	.5.15	5.12	5.11
A	6.26	5.88	5.71	6.04	6.67	6.02	5.30	5.02	5.15	4.99
ET	28.2	23.2	21.6	26.6	23.8	33.6	19.4	13.5	14.0	13.7
NF	333	273	255	315	283	406	231	154	159	155
NG	3335	2737	2550	3158	2831	4061	2313	1547	1596	1551
NDF	0	0	0	0	0	0	0	0	0	0
NDG	0	0	0	0	0	0	0	0	0	0
ET/A	4.5	4.1	3.9	4.4	3.6	5.5	4.2	2.7	2.8	2.8
NF/A	52.7	47.9	45.7	52.5	42.4	66.8	50.3	31.1	31.7	31.1
NG/A	526.8	478.6	456.8	525.3	423.9	667.7	503.0	311.1	317.1	311.4
NDF/A	0.0	0.0	0.0	0.0	0.0	0.0	0.0	0.0	0.0	0.0
NDG/A	0.0	0.0	0.0	0.0	0.0	0.0	0.0	0.0	0.0	0.0
R	0	0	0	0	0	0	0	0	0	0

Table 46: Test results for GRG2(2) (degenerate, ill-conditioned, and indefinite problems).

Class	1A	2A	3A	4A	5A	6A	7A	8A	9A	10A
FV	.23E-7	.38E-5	.13E-8	.32E-9	.21E-8	.13E-7	.23E-8	.14E-7	.11E-7	.82E-8
VC	.70E-8	.64E-5	.22E-8	.13E-9	.85E-10	.55E-9	.29E-8	.40E-7	.39E-7	.38E-7
KT	.11E-6	.39E-7	.62E-8	.70E-6	.15E-5	.17E-5	.56E-7	.57E-6	.64E-6	.85E-6
ED	7.93	9.00	9.21	7.29	6.86	6.68	8.03	7.58	6.62	7.61
ET	33.6	91.4	46.0	17.8	33.2	5.2	121.1	52.5	143.0	57.4
NF	263	403	80	107	184	98	162	150	179	204
NG	1580	4842	242	1073	1848	785	2109	2103	899	3685
NDF	263	403	80	107	184	98	162	150	179	204
NDG	850	1979	242	226	481	252	974	942	475	2007
PNS	33.3	85.2	50.0	3.3	0	0	33.3	0	26.3	11.1
FFV	.23E-3	.88E-7	.28E-1	.17E-2	-	-	.13E-7	-	.0	0.
FVC	.32E+1	.12E+3	.15E+	.0	-	-	.23E-4	-	.19E+5	.32E-1
F	2	1	6	0	0	2	0	0	0	2
PGS	43.8	0	33.3	0	0	4.2	60.0	0	0	0
GFV	-.94	-	-.29	-	-	-.17E-1	-.37	-	-	-
GVC	.72E-7	-	.27E-8	-	-	.21E-9	.43E-8	-	-	-

Table 47: Test results for VF01A (general problems).

Class	1B	2B	3B	4B	5B	6B	7B	8B	9B	10B
FV	.11E-8	.17E-9	.10E-8	.11E-11	.28E-9	.27E-9	.44E-4	.19E-8	.34E-8	.28E-9
VC	.11E-8	.55E-10	.52E-9	.23E-9	.22E-9	.42E-9	.35E-4	.34E-8	.48E-8	.42E-10
KT	.25E-6	.48E-6	.11E-5	.20E-7	.48E-6	.22E-6	.16E-3	.54E-7	.10E-7	.40E-7
ED	7.66	7.12	6.80	7.57	7.58	7.42	2.75	9.20	9.01	8.91
A	8.05	8.37	7.75	9.21	8.28	8.26	3.84	8.42	8.45	9.06
ET	31.7	29.5	38.9	7.4	20.7	39.9	62.3	37.4	27.2	27.0
NF	144	151	208	46	103	207	330	165	119	117
NG	1447	1511	2087	467	1033	2072	3303	1651	1196	1170
NDF	144	151	208	46	103	207	330	165	119	117
NDG	482	376	512	72	306	584	907	633	453	436
ET/A	3.9	3.6	5.2	0.8	2.5	4.9	16.7	4.9	3.8	3.0
NF/A	18.1	18.2	27.9	5.2	12.6	25.2	90.1	21.6	16.9	13.1
NG/A	181.4	181.8	278.6	51.7	125.7	252.0	901.4	215.8	168.5	130.7
NDF/A	18.1	18.2	27.9	5.2	12.6	25.2	90.1	21.6	16.9	13.1
NDG/A	60.0	44.8	67.6	8.1	37.3	71.0	239.9	83.2	64.3	48.8
R	0	1	1	0	0	0	0	0	0	0

Table 48: Test results for VF01A (degenerate, ill-conditioned, and indefinite problems).

Class	1A	2A	3A	4A	5A	6A	7A	8A	9A	10A
FV	.99E-7	.41E-6	.50E-8	.11E-4	.15E-4	.15E-6	-	.90E-7	.54E-6	.51E-6
VC	.25E-6	.78E-6	.17E-7	.34E-4	.38E-4	.12E-7	-	.32E-6	.15E-5	.25E-5
KT	.83E-6	.76E-6	.96E-7	.91E-3	.86E-3	.26E-5	-	.38E-5	.51E-4	.55E-4
ED	6.91	8.25	8.21	3.30	3.33	6.15	-	6.76	3.44	5.68
ET	36.5	129.3	19.5	54.6	53.9	7.7	-	63.4	171.7	72.4
NF	469	644	57	226	194	168	-	225	364	352
NG	2813	7731	172	2261	1948	1347	-	3155	1822	6352
NDF	154	147	28	113	113	62	-	83	121	117
NDG	929	1764	85	1132	1138	502	-	1170	608	2123
PNS	29.2	87.8	72.2	3.3	0	0	93.3	0	20.0	26.7
FFV	.12E-3	.24E-2	.0	.34E-2	-	-	.29E-9	-	.0	.26E-2
FVC	.82	.67E+2	.20E+12	.0	-	-	.54E-1	-	.26E+2	.86E-5
F	2	4	4	0	0	0	0	0	0	0
PGS	41.2	0	0	10.3	0	10.0	100.0	0	0	0
GFV	-.21	-	-	-.15E-2	-	-.17E-1	-.87E-1	-	0	-
GVC	.12E-5	-	-	.22E-3	-	.60E-7	.15E-5	-	-	-

Table 49: Test results for LPNLP (general problems).

Class	1B	2B	3B	4B	5B	6B	7B	8B	9B	10B
FV	.21E-4	.21E-5	.14E-4	.43E-11	.61E-5	.14E-4	.19E-3	.86E-8	.17E-8	.62E-8
VC	.40E-4	.35E-5	.36E-4	.99E-8	.19E-4	.35E-4	.22E-3	.44E-8	.11E-7	.14E-7
KT	.32E-2	.27E-3	.23E-2	.13E-6	.19E-2	.12E-2	.16E-2	.18E-7	.17E-7	.39E-7
ED	3.09	3.60	2.82	5.76	3.87	3.32	2.09	8.02	8.12	7.66
A	3.67	4.58	3.69	8.00	4.13	3.89	3.07	8.05	8.15	7.79
ET	97.3	66.9	95.9	9.6	76.2	86.2	116.5	24.4	26.2	24.6
NF	386	272	410	41	255	425	631	119	137	124
NG	3863	2722	4103	412	2555	4250	6313	1197	1372	1240
NDF	187	127	179	17	160	157	187	42	43	42
NDG	1870	1278	1793	173	1607	1571	1878	423	435	421
ET/A	26.9	16.5	26.9	1.2	19.3	23.1	37.9	3.1	3.3	3.1
NF/A	107.2	67.3	116.4	5.1	64.3	113.8	206.1	15.3	17.1	15.8
NG/A	1072.0	672.8	1164.2	50.8	643.2	1137.6	2060.6	152.9	171.2	158.0
NDF/A	51.8	31.4	49.9	2.1	40.9	42.3	61.1	5.3	5.4	5.4
NDG/A	518.0	313.6	499.8	21.5	408.5	422.8	611.4	53.4	53.9	53.8
R	0	0	0	0	0	0	0	0	0	0

Table 50: Test results for LPNLP (degenerate, ill-conditioned, and indefinite problems).

Class	1A	2A	3A	4A	5A	6A	7A	8A	9A	10A
FV	.49E-4	-	.34E-6	.85E-4	.13E-3	.12E-3	-	.18E-3	.12E-3	.82E-4
VC	.61E-5	-	.54E-6	.45E-5	.12E-7	.31E-5	-	.11E-4	.26E-4	.55E-5
KT	.49E-2	-	.81E-4	.90E-2	.70E-2	.17E-1	-	.35E-1	.56E-2	.24E-1
ED	3.63	-	5.40	2.81	2.64	3.01	-	2.87	2.30	3.27
ET	48.0	-	27.1	23.3	27.3	5.4	-	69.2	138.6	54.3
NF	353	-	47	59	65	55	-	117	115	135
NG	2118	-	143	591	652	445	-	1640	578	2434
NDF	350	-	46	55	61	54	-	115	112	132
NDG	2105	-	140	554	618	437	-	1621	562	2376
PNS	16.7	-	36.7	40.0	53.3	85.2	86.7	66.7	55.6	46.7
FFV	.11E-3	-	.83E-9	.21E-1	.24E-2	.10E-2	.94E-5	.46E-2	.45E-2	.41E-2
FVC	.20	-	.29E-1	.13E-5	.26E-5	.17E-5	.36E-2	.14E-4	.43E-6	.81E-4
F	0	10	0	0	0	1	0	0	2	0
PGS	60.0	-	73.7	5.6	0	25.0	100.0	0	0	0
GFV	-.11E+1	-	-.39E+1	-.74E-4	-	-.17E-1	-.18	-	-	-
GVC	.76E-5	-	.32E-5	.47E-3	-	.71E-6	.17E-4	-	-	-

Table 51: Test results for SALQDR (general problems).

Class	1B	2B	3B	4B	5B	6B	7B	8B	9B	10B
FV	.27E-3	.15E-3	.61E-4	.13E-4	.96E-4	.33E-3	.58E-3	.52E-3	.10E-3	.12E-3
VC	.10E-5	.14E-4	.21E-4	.20E-4	.15E-5	.14E-4	.77E-4	.34E-8	.83E-7	.38E-7
KT	.26E-1	.12E-1	.58E-2	.19E-2	.13E-1	.23E-1	.23E-1	.98E-2	.41E-2	.36E-2
ED	2.38	2.50	2.43	2.62	3.09	2.40	2.13	4.14	4.69	4.60
A	3.38	3.28	3.39	3.73	3.71	3.09	2.80	4.48	4.54	4.59
ET	23.6	36.5	28.7	25.5	21.0	26.3	28.7	19.0	21.3	21.2
NF	57	85	69	59	48	62	68	41	46	47
NG	570	856	693	598	487	628	683	415	461	470
NDF	50	79	63	55	44	56	60	40	45	45
NDG	500	798	630	553	447	566	606	405	451	455
ET/A	7.4	11.1	8.7	7.3	6.0	8.7	10.2	4.3	4.8	4.6
NF/A	17.6	26.1	21.0	17.2	13.9	21.0	24.4	9.4	10.3	10.3
NG/A	175.8	261.3	209.5	171.9	138.7	210.0	244.2	94.2	103.0	102.8
NDF/A	15.8	24.3	19.1	15.9	12.8	18.8	21.6	9.2	10.1	9.9
NDG/A	155.7	243.3	190.5	158.9	127.8	188.3	216.1	92.0	100.8	99.2
R	2	1	1	0	0	0	0	1	1	0

Table 52: Test results for SALQDR (degenerate, ill-conditioned, and indefinite problems).

Class	1A	2A	3A	4A	5A	6A	7A	8A	9A	10A
FV	.53E-4	–	.91E-7	.71E-4	.23E-3	.17E-3	–	.19E-3	.17E-3	.22E-4
VC	.45E-5	–	.38E-6	.40E-5	.81E-8	.29E-6	–	.15E-4	.75E-5	.13E-5
KT	.89E-2	–	.41E-4	.61E-2	.12E-1	.12E-1	–	.37E-1	.11E-1	.12E-1
ED	3.39	–	5.80	2.91	2.59	3.10	–	2.71	2.37	3.54
ET	128.2	–	49.5	43.2	43.4	20.4	–	191.1	160.9	388.2
NF	2798	–	493	509	639	866	–	1910	1658	3265
NG	16645	–	1303	4430	5688	6253	–	25358	7605	55295
NDF	0	–	0	0	0	0	–	0	0	0
NDG	0	–	0	0	0	0	–	0	0	0
PNS	20.0	–	33.3	53.3	46.7	59.3	93.3	60.0	44.4	66.7
FFV	.17E-4	–	.14E-8	.89E-2	.33E-2	.60E-2	.37E-2	.36E-2	.42E-2	.37E-2
FVC	.73	–	.24E-1	.99E-7	.54E-5	.52E-5	.11E-3	.31E-4	.29E-6	.28E-3
F	0	10	0	0	0	1	0	0	2	2
PGS	62.5	–	75.0	7.1	0	0	100.0	0	0	0
GFV	-.11E+1	–	-.46E+1	-.14E-2	–	–	-.64	–	–	–
GVC	.53E-5	–	.13E-4	.19E-3	–	–	.30E-6	–	–	–

Table 53: Test results for SAIQDF (general problems).

Class	1B	2B	3B	4B	5B	6B	7B	8B	9B	10B
FV	.24E-3	.17E-3	.38E-4	.67E-5	.21E-3	.11E-2	.88E-3	.48E-5	.89E-5	.56E-5
VC	.44E-5	.87E-5	.87E-6	.23E-5	.80E-5	.23E-6	.96E-4	.44E-7	.72E-6	.23E-8
KT	.17E-1	.13E-1	.68E-2	.23E-2	.14E-1	.41E-1	.38E-1	.22E-3	.86E-3	.30E-3
ED	2.45	2.46	2.68	3.03	2.91	2.38	2.13	5.57	5.17	5.46
A	3.30	3.28	3.83	4.12	3.39	3.34	2.66	5.48	4.86	5.72
ET	44.4	42.9	37.8	43.1	35.3	38.3	41.5	36.6	37.5	41.3
NF	646	617	563	633	538	573	610	544	533	582
NG	5765	5445	4908	5575	4642	5051	5452	4744	4642	5142
NDF	0	0	0	0	0	0	0	0	0	0
NDG	0	0	0	0	0	0	0	0	0	0
ET/A	14.0	13.4	10.2	10.9	10.7	12.5	16.2	7.0	8.0	8.0
NF/A	203.4	192.3	151.7	161.2	163.6	185.6	237.3	104.2	115.4	113.0
NG/A	1814.3	1696.9	1321.2	1419.6	1412.9	1640.6	2127.5	899.8	994.2	997.3
NDF/A	.0	.0	.0	.0	.0	.0	.0	.0	.0	.0
NDG/A	.0	.0	.0	.0	.0	.0	.0	.0	.0	.0
R	0	1	1	0	0	0	0	1	2	2

Table 54: Test results for SALQDF (degenerate, ill-conditioned, and indefinite problems).

Class	1A	2A	3A	4A	5A	6A	7A	8A	9A	10A
FV	.52E-5	–	.26E-5	.13E-5	.14E-5	.37E-9	–	.43E-9	.24E-5	.16E-8
VC	.15E-4	–	.21E-4	.86E-6	.20E-5	.39E-10	–	.21E-9	.19E-5	.53E-8
KT	.51E-3	–	.17E-5	.31E-3	.47E-3	.48E-7	–	.28E-6	.76E-3	.37E-6
ED	4.61	–	6.62	4.06	3.96	8.26	–	7.81	2.70	7.78
ET	147.6	–	66.4	79.5	110.5	22.3	–	285.0	409.9	288.7
NF	429	–	24	35	46	61	–	74	28	70
NG	7235	–	356	2054	2623	2579	–	7347	1686	16690
NDF	1205	–	118	205	262	322	–	524	337	927
NDG	7235	–	356	2054	2623	2579	–	7347	1686	16690
PNS	20.0	–	20.0	10.0	0	11.1	88.9	0	0	20.0
FPV	.26E-5	–	.0	.17E-2	–	.18E-2	.21E-3	–	–	.43E-2
FVC	.14E-1	–	.14E+1	.14E-11	–	.30E-4	.20E-2	–	–	.17E-3
F	0	10	0	0	0	1	2	1	4	0
PGS	62.5	–	79.2	0	0	12.5	100.0	0	0	0
GFV	-.11E+1	–	-.29E+1	–	–	-.17E-1	-.87E-1	–	–	–
GVC	.50E-6	–	.23E-4	–	–	.62E-10	.47E-6	–	–	–

Table 55: Test results for SALMNF (general problems).

Class	1B	2B	3B	4B	5B	6B	7B	8B	9B	10B
FV	.34E-4	.29E-5	.60E-5	.61E-9	.31E-5	.24E-5	.16E-3	.22E-11	.62E-11	.34E-11
VC	.43E-7	.66E-7	.50E-7	.59E-8	.50E-6	.23E-6	.78E-6	.13E-11	.13E-11	.13E-11
KT	.38E-2	.86E-3	.96E-3	.26E-5	.11E-2	.92E-3	.94E-2	.16E-10	.71E-10	.28E-10
ED	3.16	3.69	3.66	5.65	4.00	3.58	2.04	10.92	10.53	10.88
A	4.35	4.87	4.80	7.17	4.69	4.72	3.49	11.31	10.94	11.20
ET	122.8	109.0	110.2	39.7	92.5	104.1	80.8	56.7	55.7	55.0
NF	58	45	49	14	39	40	48	19	19	19
NG	2805	2432	2511	897	2157	2443	1867	1318	1300	1286
NDF	280	243	251	89	216	244	186	131	130	128
NDG	2805	2432	2511	897	2157	2443	1867	1318	1300	1286
ET/A	28.0	23.3	24.4	6.5	20.4	23.4	23.9	5.0	5.2	4.9
NF/A	12.9	9.9	11.0	2.4	8.9	8.9	14.7	1.7	1.8	1.7
NG/A	638.0	520.3	553.9	146.6	476.2	547.9	551.3	117.0	120.6	115.7
NDF/A	63.8	52.0	55.4	14.7	47.6	54.8	55.1	11.7	12.1	11.6
NDG/A	638.0	520.3	553.9	146.6	476.2	547.9	551.3	117.0	120.6	115.7
R	0	0	0	0	0	0	0	0	0	0

Table 56: Test results for SALMNF (degenerate, ill-conditioned, and indefinite problems).

Class	1A	2A	3A	4A	5A	6A	7A	8A	9A	10A
FV	.16E-3	–	.32E-7	.46E-6	.23E-6	.19E-5	–	.21E-5	.49E-5	–
VC	.12E-4	–	.12E-6	.63E-9	.18E-8	.56E-7	–	.22E-5	.13E-6	–
KT	.27E-1	–	.10E-6	.30E-3	.20E-3	.45E-3	–	.82E-3	.13E-2	–
ED	2.87	–	8.40	4.78	4.81	4.37	–	4.42	3.96	–
ET	76.5	–	56.4	94.4	97.2	26.7	–	196.8	165.8	–
NF	1462	–	368	799	983	921	–	1363	957	–
NG	8775	–	1104	7991	9530	7369	–	19094	4786	–
NDF	106	–	31	54	64	58	–	82	58	–
NDG	639	–	93	543	652	467	–	1160	292	–
PNS	83.3	100.0	76.7	40.0	26.7	37.5	–	13.3	65.6	–
FPV	.45E-5	.65E-4	.91E-7	.42E-1	.48E-2	.45E-1	–	.12E-2	.25E-3	–
FVC	.66E-2	.33E-1	.30	.25E-8	.12E-8	.14E-6	–	.11E-3	.19E-5	–
F	0	0	0	0	0	2	5	0	2	5
PGS	60.0	0	28.6	0	0	0	–	0	0	–
GFV	-.73E+1	–	-.82E-2	–	–	–	–	–	–	–
GVC	.19E-3	–	.11E-3	–	–	–	–	–	–	–

Table 57: Test results for CONMIN (general problems).

Class	1B	2B	3B	4B	5B	6B	7B	8B	9B	10B
FV	.89E-7	.25E-7	.31E-7	.31E-7	.63E-7	.18E-7	.17E-6	.81E-8	.44E-8	.37E-8
VC	.84E-8	.61E-8	.85E-8	.16E-10	.17E-6	.93E-8	.55E-8	.23E-9	.17E-9	.43E-8
KT	.13E-3	.12E-3	.15E-3	.16E-3	.97E-4	.21E-4	.34E-4	.13E-5	.20E-6	.17E-6
ED	4.88	4.35	4.62	4.28	5.27	5.59	3.23	7.58	7.80	7.72
A	5.97	6.02	6.00	6.59	5.81	6.52	5.68	7.55	8.16	7.82
ET	58.8	60.8	63.8	66.7	49.9	61.6	117.4	29.3	31.4	30.6
NF	458	505	535	574	401	492	1056	217	233	227
NG	4587	5057	5355	5743	4012	4923	10560	2176	2332	2275
NDF	40	43	45	48	38	42	63	18	20	19
NDG	405	431	455	480	381	420	638	187	202	196
ET/A	9.9	10.0	10.6	10.3	8.6	9.5	37.2	3.9	4.0	4.1
NF/A	77.5	83.5	90.2	89.1	69.1	76.2	347.0	29.3	29.8	30.9
NG/A	774.5	834.8	902.3	890.9	691.4	761.6	3460.1	292.9	297.9	308.6
NDF/A	6.8	7.1	7.6	7.4	6.6	6.5	18.3	2.5	2.5	2.6
NDG/A	68.3	71.3	76.4	74.3	65.8	65.3	183.4	24.6	25.3	26.1
R	0	0	0	0	0	0	0	0	0	0

Table 58: Test results for CONMIN (degenerate, ill-conditioned, and indefinite problems).

Class	1A	2A	3A	4A	5A	6A	7A	8A	9A	10A
FV	.27E-4	.14E-3	.21E-5	.91E-6	.19E-5	.90E-5	.55E-6	.49E-6	.76E-6	.14E-4
VC	.40E-5	.31E-3	.52E-5	.11E-5	.39E-5	.57E-6	.64E-5	.17E-5	.12E-5	.43E-6
KT	.13E-4	.73E-6	.14E-5	.20E-3	.22E-3	.66E-4	.41E-3	.36E-4	.40E-4	.12E-3
ED	5.70	7.39	6.86	4.31	4.18	4.58	4.10	5.79	4.98	5.18
ET	26.1	82.8	53.3	62.8	45.7	17.7	220.8	102.1	123.8	173.5
NF	416	604	317	458	372	617	691	601	543	1070
NG	2779	7788	1062	5228	4238	5533	10157	9515	3049	21733
NDF	45	43	35	59	48	71	87	75	65	133
NDG	270	516	107	593	485	575	1131	1062	328	2410
PNS	33.3	96.7	3.7	3.3	0	6.7	25.0	0	75.0	20.0
FPV	.59E-6	.21E-1	.0	.17E-2	-	.0	.0	-	.80E-5	.0
FVC	.17E-1	.30E-9	.10E-2	.0	-	.11E-2	.53E-2	-	.29E-3	.62E-2
P	1	0	1	0	0	0	1	0	1	0
PGS	66.7	0	76.9	0	0	10.7	66.7	0	0	16.7
GPV	-.14E+1	-	-.65E+3	-	-	-.17E-1	-.32	-	-	-.15E-2
GVC	.16E-4	-	.21E-4	-	-	.39E-10	.10E-3	-	-	.15E-4

Table 59: Test results for BIAS(1) (general problems).

Class	1B	2B	3B	4B	5B	6B	7B	8B	9B	10B
FV	.43E-4	.35E-5	.11E-4	.13E-10	.56E-5	.67E-5	.76E-5	.18E-4	.28E-5	.19E-6
VC	.58E-4	.93E-5	.14E-4	.35E-7	.12E-4	.18E-4	.16E-4	.23E-5	.57E-8	.12E-6
KT	.34E-2	.30E-3	.11E-2	.48E-6	.13E-2	.78E-3	.35E-4	.35E-4	.37E-5	.54E-6
ED	3.00	3.49	3.26	5.42	4.06	3.60	2.71	6.21	6.42	6.85
A	3.52	4.38	4.01	7.53	4.29	4.16	4.27	5.26	6.41	6.69
ET	73.2	83.1	81.0	16.6	75.6	66.2	106.1	29.2	28.0	24.4
NF	525	612	597	139	542	508	865	249	240	207
NG	6146	7107	6982	1576	6441	5856	9777	2794	2681	2301
NDF	83	92	94	16	91	72	108	27	25	20
NDG	830	920	945	165	912	722	1083	272	257	207
ET/A	21.8	20.5	21.0	2.2	18.1	16.8	26.3	5.7	4.6	3.6
NF/A	155.9	150.1	154.8	18.2	129.7	128.7	213.9	48.9	39.8	30.6
NG/A	1823.8	1742.6	1808.3	207.0	1539.2	1483.9	2419.6	547.7	443.0	339.6
NDF/A	24.7	22.8	24.4	2.2	21.9	18.4	26.9	5.4	4.2	3.1
NDG/A	246.6	227.6	244.4	22.0	218.8	183.7	268.9	53.9	42.3	30.8
R	0	0	0	0	0	0	0	1	1	1

Table 60: Test results for BIAS(1) (degenerate, ill-conditioned, and indefinite problems).

Class	1A	2A	3A	4A	5A	6A	7A	8A	9A	10A
FV	.49E-5	.15E-3	.73E-5	.54E-5	.26E-5	.25E-4	.17E-4	.33E-4	.40E-3	.59E-4
VC	.17E-5	.39E-3	.13E-4	.45E-6	.50E-5	.48E-5	.52E-4	.70E-5	.18E-4	.48E-4
KT	.81E-4	.17E-5	.67E-4	.10E-2	.33E-3	.22E-2	.43E-2	.14E-1	.18E-1	.13E-1
ED	5.15	7.16	5.19	3.63	4.03	3.17	3.06	3.15	2.44	3.34
ET	39.0	77.9	70.0	108.7	74.3	46.5	376.4	247.8	334.3	514.8
NF	837	681	658	1269	1082	2216	2417	2500	3471	4457
NG	5036	8196	1978	12730	10851	17756	31460	35062	17368	80308
NDF	0	0	0	0	0	0	0	0	0	0
NDG	0	0	0	0	0	0	0	0	0	0
PNS	22.2	95.3	8.3	10.0	3.3	33.3	22.2	0	66.7	33.3
FFV	.19E-2	.22E-1	.56E+2	.14E-5	.0	.81E-11	.64E-7	-	.64E-3	.37E-8
FVC	.22E-1	.20E-8	.68E-5	.70E-4	.12E-2	.42E-2	.16E-2	-	.49E-3	.82E-2
F	4	0	6	0	0	0	2	0	1	0
PGS	57.1	50.0	54.5	0	0	15.0	57.1	0	25.0	10.0
GFV	-.56	-.22E-1	-.85E+2	-	1	-.17E-1	-.24	-	-.20E+2	-.11E-2
GVC	.27E-4	.20E-8	.59E-5	-	-	.30E-5	.77E-4	-	.15E-3	.23E-3

Table 61: Test results for BIAS(2) (general problems).

Class	1B	2B	3B	4B	5B	6B	7B	8B	9B	10B
FV	.37E-4	.29E-4	.34E-4	.17E-6	.74E-6	.22E-4	.29E-4	.11E-4	.65E-5	.33E-5
VC	.87E-5	.36E-4	.55E-4	.38E-7	.54E-5	.15E-4	.25E-4	.27E-5	.11E-8	.51E-6
KT	.32E-2	.20E-2	.27E-2	.15E-3	.79E-3	.18E-2	.70E-3	.15E-3	.33E-3	.16E-3
ED	2.86	2.86	2.78	3.46	4.04	3.33	2.49	4.64	4.69	4.94
A	3.71	3.63	3.52	5.37	4.63	3.89	3.70	4.67	5.58	5.13
ET	138.1	164.7	159.9	96.5	131.8	140.6	173.7	67.0	66.6	64.1
NF	1737	2084	2024	1224	1640	1751	2187	838	826	795
NG	17416	20886	20275	12271	16452	17544	21904	8408	8287	7978
NDF	0	0	0	0	0	0	0	0	0	0
NDG	0	0	0	0	0	0	0	0	0	0
ET/A	39.1	47.7	47.3	20.3	29.3	36.9	49.2	14.4	12.8	12.7
NF/A	491.6	603.8	599.0	257.1	365.3	456.0	620.5	179.9	159.2	158.2
NG/A	4927.6	6049.0	5999.3	2576.9	3663.5	4606.9	6213.3	1803.7	1595.6	1585.7
NDF/A	0.0	0.0	0.0	0.0	0.0	0.0	0.0	0.0	0.0	0.0
NDG/A	0.0	0.0	0.0	0.0	0.0	0.0	0.0	0.0	0.0	0.0
R	0	0	0	0	0	1	2	1	0	1

Table 62: Test results for BIAS(2) (degenerate, ill-conditioned, and indefinite problems).

221

Class	1A	2A	3A	4A	5A	6A	7A	8A	9A	10A
FV	.22E-3	-	.18E-8	.31E-9	.15E-9	.15E-8	.51E-9	.29E-9	.20E-8	.16E-7
VC	.12E-4	-	.26E-8	.16E-9	.43E-9	.13E-9	.25E-8	.12E-8	.18E-8	.20E-8
KT	.24E-1	-	.13E-5	.80E-5	.47E-5	.97E-5	.36E-5	.12E-4	.71E-5	.95E-4
ED	3.41	-	6.82	6.35	6.48	6.22	6.47	6.30	5.10	5.65
ET	35.5	-	104.0	79.6	77.0	13.7	438.8	116.0	328.3	122.3
NF	517	-	516	510	499	333	1056	520	929	594
NG	3102	-	1548	5109	4995	2671	13728	7283	4648	10707
NDF	117	-	101	104	108	74	244	118	199	138
NDG	702	-	304	1043	1082	594	3172	1661	998	2496
PNS	83.3	100.0	25.9	10.0	0	0	65.6	0	0	46.7
FFV	.47E-3	.74	.18E-7	.60E-2	-	-	.65E-8	-	-	.30
FVC	.47E+1	.12E-1	.15E+6	.0	-	-	.11E-3	-	-	.27E-3
F	0	1	1	0	0	0	2	0	1	0
PGS	80.0	0	80.0	0	0	10.0	75.0	0	0	12.5
GFV	-.22E+2	-	-.35E+3	-	-	-.17E-1	-.87E-1	-	-	-.16E-2
GVC	.24E-4	-	.94E-5	-	-	.41E-7	.10E-3	-	-	.55E-5

Table 63: Test results for FUNMIN (general problems).

Class	1B	2B	3B	4B	5B	6B	7B	8B	9B	10B
FV	.92E-8	.12E-9	.81E-9	.12E-11	.97E-10	.11E-9	.19E-6	.13E-10	.23E-10	.30E-10
VC	.77E-8	.50E-10	.67E-10	.13E-10	.20E-9	.11E-9	.43E-9	.18E-10	.12E-10	.21E-10
KT	.25E-4	.44E-5	.14E-4	.60E-7	.24E-5	.32E-5	.10E-3	.81E-7	.19E-6	.17E-6
ED	5.38	5.91	5.49	7.76	6.86	6.37	2.89	8.68	8.46	8.43
A	6.53	7.87	7.15	9.45	8.05	7.95	5.74	9.35	9.18	9.10
ET	72.9	82.6	73.2	23.1	61.6	93.4	118.8	28.6	28.0	27.3
NF	451	514	468	141	391	605	766	175	170	162
NG	4515	5142	4681	1412	3917	6058	7668	1751	1701	1628
NDF	99	111	101	32	83	129	162	40	39	37
NDG	992	1116	1010	321	838	1292	1625	403	395	372
ET/A	12.0	10.8	10.8	2.5	7.7	11.7	21.7	3.1	3.0	3.0
NF/A	74.5	67.3	69.4	15.4	48.9	75.6	139.7	18.9	18.5	18.0
NG/A	745.0	672.6	693.5	154.2	488.9	756.4	1396.7	188.7	184.6	179.8
NDF/A	16.4	14.6	15.0	3.5	10.5	16.1	29.8	4.3	4.3	4.1
NDG/A	164.1	146.1	150.2	35.1	104.7	160.9	297.5	43.4	42.8	41.0
R	0	0	0	0	0	0	0	0	0	0

Table 64: Test results for FUNMIN (degenerate, ill-conditioned, and indefinite problems).

Class	1A	2A	3A	4A	5A	6A	7A	8A	9A	10A
FV	.61E-4	–	.13E-5	.30E-5	.27E-4	.40E-6	–	.73E-3	.39E-4	.26E-3
VC	.45E-6	–	.68E-6	.24E-11	.16E-11	.42E-10	–	.0	.19E-8	.22E-7
KT	.60E-3	–	.16E-2	.13E-2	.45E-2	.89E-3	–	.69E-1	.73E-3	.27E-1
ED	4.66	–	3.71	4.27	3.28	3.98	–	2.67	3.01	3.12
ET	171.0	–	35.7	48.5	82.0	13.8	–	203.9	167.2	183.9
NF	344	–	58	105	163	131	–	291	122	332
NG	2067	–	174	1052	1632	1052	–	4074	611	5985
NDF	344	–	58	105	163	131	–	291	122	332
NDG	2067	–	174	1052	1632	1052	–	4074	611	5985
PNS	25.9	100.0	33.3	0	16.7	10.0	–	33.3	8.3	66.7
FFV	.20E-1	.97E-7	.20E-1	–	.74E-2	.72E-1	–	.75E-2	.52E-2	.52E-2
FVC	.72E-6	.19E+5	.16E-5	–	.0	.0	–	.78E-7	.11E-7	.13E-9
F	1	9	4	0	0	0	5	4	1	0
PGS	80.0	0	83.3	0	0	3.7	–	0	0	20.0
GFV	-.15E+1	–	-.43	–	–	-.17E-1	–	–	–	-.89E-3
GVC	.37E-6	–	.19E-5	–	–	.23E-7	–	–	–	.0

Table 65: Test results for GAPFPR (general problems).

Class	1B	2B	3B	4B	5B	6B	7B	8B	9B	10B
FV	.80E-4	.4E-4	.18E-4	.70E-4	.84E-4	.53E-4	.25E-3	.25E-2	.12E-2	.27E-2
VC	.97E-11	.24E-11	.19E-11	.37E-11	.0	.0	.0	.46E-10	.46E-10	.67E-10
KT	.76E-2	.32E-2	.41E-2	.92E-2	.80E-2	.21E-2	.28E-2	.18E-2	.61E-2	.43E-3
ED	2.78	2.99	3.02	2.67	3.21	3.06	2.02	3.24	3.50	3.35
A	5.00	5.49	5.47	5.07	5.34	5.50	5.05	4.73	4.74	4.86
ET	128.0	93.4	102.0	112.8	88.1	65.0	81.3	116.2	112.2	128.4
NF	243	181	195	215	172	123	154	214	208	238
NG	2435	1810	1952	2150	1724	1237	1545	2140	2080	2382
NDF	243	181	195	215	172	123	154	214	208	238
NDG	2435	1810	1952	2150	1724	1237	1545	2140	2080	2382
ET/A	27.4	16.9	19.0	23.5	17.0	11.9	16.2	25.6	23.3	26.4
NF/A	52.1	32.7	36.5	44.6	33.2	22.6	30.8	47.4	43.4	48.9
NG/A	521.5	327.2	364.5	446.1	332.3	226.1	307.6	474.3	432.5	488.7
NDF/A	52.1	32.7	36.5	44.6	33.2	22.6	30.8	47.4	43.4	48.9
NDG/A	521.5	327.2	364.5	446.1	332.3	226.1	307.6	474.3	432.5	488.7
R	0	0	0	0	1	1	0	3	1	2

Table 66: Test results for GAPFPR (degenerate, ill-conditioned, and indefinite problems).

Class	1A	2A	3A	4A	5A	6A	7A	8A	9A	10A
FV	.15E-4	.13E-5	.10E-5	.25E-5	.72E-6	.11E-5	–	.42E-6	.31E-5	.50E-6
VC	.55E-5	.49E-6	.23E-5	.63E-6	.14E-5	.24E-6	–	.72E-6	.11E-5	.49E-6
KT	.74E-2	.25E-5	.33E-5	.23E-3	.11E-3	.27E-5	–	.39E-5	.69E-4	.17E-5
ED	3.01	6.46	6.21	3.99	4.20	5.57	–	5.96	3.90	6.66
ET	43.4	77.6	28.8	66.4	69.9	10.6	–	73.3	214.3	61.4
NF	141	223	66	156	148	121	–	164	209	150
NG	1112	2217	377	1537	1515	912	–	2030	1443	1888
NDF	141	223	66	156	148	121	–	164	209	150
NDG	1112	2217	377	1537	1515	912	–	2030	1443	1888
PNS	0	52.4	50.0	0	6.7	6.7	–	0	0	0
FPV	–	.30E-4	.0	–	.61E-1	.83E-2	–	–	–	–
FVC	–	.37E-4	.59E+8	–	.90E-9	.55E-5	–	–	–	–
F	9	3	4	5	0	0	5	0	2	2
PGS	0	0	0	0	0	7.1	–	0	0	0
GFV	–	–	–	–	–	-.17E-1	–	–	–	–
GVC	–	–	–	–	–	.59E-9	–	–	–	–

__Table 67:__ Test results for GAPFQL (general problems).

Class	1B	2B	3B	4B	5B	6B	7B	8B	9B	10B
FV	.29E-5	.31E-4	.30E-4	.15E-9	.79E-6	.10E-2	.13E-2	.12E-6	.37E-5	.18E-5
VC	.40E-5	.26E-7	.27E-7	.13E-5	.26E-8	.0	.92E-5	.12E-7	.37E-5	.11E-6
KT	.13E-2	.44E-2	.23E-2	.18E-5	.23E-3	.27E-1	.35E-1	.47E-6	.33E-5	.34E-5
ED	3.78	2.89	3.32	4.68	4.50	2.51	2.13	7.16	6.56	6.36
A	4.40	4.34	4.51	6.54	5.71	4.77	2.87	7.08	5.73	6.13
ET	121.2	103.0	114.0	13.5	43.5	185.2	131.4	40.9	77.1	77.3
NF	239	207	233	27	92	390	278	82	155	156
NG	2396	2076	2336	273	920	3900	2785	828	1558	1563
NDF	239	207	233	27	92	390	278	82	155	156
NDG	2396	2076	2336	273	920	3900	2785	828	1558	1563
ET/A	27.4	25.3	25.2	2.0	6.9	38.5	45.9	6.1	17.7	14.9
NF/A	54.4	51.5	51.9	4.1	14.5	80.8	97.3	12.3	35.7	30.3
NG/A	544.2	515.1	518.8	40.5	145.1	807.8	972.6	123.3	357.4	302.5
NDF/A	54.4	51.5	51.9	4.1	14.5	80.8	97.3	12.3	35.7	30.3
NDG/A	544.2	515.1	518.8	40.5	145.1	807.8	972.6	123.3	357.4	302.5
R	1	0	2	0	1	1	4	0	0	2

Table 68: Test result for GAPFQL (degenerate, ill-conditioned, and indefinite problems)

Class	1A	2A	3A	4A	5A	6A	7A	8A	9A	10A
FV	.16E-5	.11E-5	.16E-5	.39E-5	.37E-5	.36E-5	–	.72E-6	.16E-5	.11E-5
VC	.99E-6	.23E-5	.36E-5	.65E-8	.38E-7	.10E-6	–	.29E-5	.39E-5	.44E-5
KT	.36E-5	.32E-5	.15E-5	.30E-3	.60E-3	.15E-4	–	.28E-4	.66E-4	.56E-4
ED	6.16	7.93	6.48	4.08	3.83	5.06	–	5.82	3.65	5.59
ET	53.6	157.8	49.6	33.5	47.2	17.3	–	113.9	127.7	145.2
NF	446	614	79	109	148	153	–	225	132	343
NG	4126	10337	447	2003	2749	2396	–	5818	1243	10805
NDF	236	239	66	82	115	135	–	179	109	249
NDG	692	939	198	272	484	415	–	1408	347	2886
PNS	8.5	28.6	20.0	0	0	0	50.0	0	0	20.0
FFV	.51E-4	.84E+2	.0	–	–	–	.13	–	–	.24E-2
FVC	.15E+1	.18E-9	.48	–	–	–	.11E-5	–	–	.13E-5
F	1	3	0	0	0	0	3	0	1	0
PGS	59.1	6.7	87.5	0	0	10.0	100.0	0	0	0
GFV	-.69	-.20E+1	-.36E+1	–	–	-.17E-1	-.12E+1	–	–	–
GVC	.91E-5	.11E-5	.19E-5	–	–	.26E-8	.66E-7	–	–	–

Table 69: Test results for ACDPAC (general problems).

Class	1B	2B	3B	4B	5B	6B	7B	8B	9B	10B
FV	.26E-5	.17E-5	.13E-5	.25E-5	.40E-5	.16E-5	.12E-5	.14E-5	.16E-5	.76E-7
VC	.18E-5	.39E-8	.95E-6	.54E-10	.71E-9	.24E-7	.37E-8	.14E-9	.13E-6	.19E-7
KT	.15E-5	.92E-3	.51E-3	.15E-2	.13E-2	.46E-3	.20E-4	.12E-5	.29E-5	.80E-6
ED	4.72	3.46	3.78	3.23	4.02	3.79	2.60	6.24	6.24	6.49
A	5.46	5.17	4.75	5.48	5.36	5.13	5.42	6.96	6.11	6.86
ET	19.4	40.6	39.2	45.0	36.5	41.7	56.6	29.0	25.3	28.6
NF	57	125	130	147	116	130	189	89	80	101
NG	1033	2332	2317	2673	2122	2391	3381	1636	1445	1710
NDF	43	98	93	110	86	99	137	68	59	65
NDG	201	407	379	426	364	408	556	275	237	254
ET/A	3.7	8.1	8.3	8.2	7.0	8.2	10.6	4.2	4.2	4.3
NF/A	11.1	24.9	27.6	26.8	22.4	25.7	35.5	12.9	13.5	15.1
NG/A	200.0	462.7	490.7	488.8	404.9	469.9	633.1	236.2	242.3	255.9
NDF/A	8.3	19.6	19.7	20.3	16.3	19.5	25.8	9.9	9.9	9.8
NDG/A	38.9	81.0	80.6	78.4	69.1	79.6	103.8	40.1	39.1	37.6
R	0	0	0	0	0	0	0	0	0	0

Table 70: Test results for ACDPAC (degenerate, ill-conditioned, and indefinite problems).

Class	1A	2A	3A	4A	5A	6A	7A	8A	9A	10A
FV	.39E-4	.12E-3	.16E-7	.47E-5	.82E-5	.35E-5	.18E-4	.78E-3	.14E-3	-
VC	.39E-5	.48E-6	.41E-7	.55E-7	.19E-8	.87E-6	.51E-4	.0	.69E-5	-
KT	.30E-2	.37E-4	.36E-6	.76E-3	.15E-2	.68E-3	.17E-2	.38E-1	.72E-2	-
ED	3.72	6.44	7.63	3.70	3.33	3.87	3.47	2.68	2.67	-
ET	66.5	232.2	131.1	109.4	99.3	27.3	286.2	150.2	233.7	-
NF	1002	1516	652	663	596	643	665	694	602	-
NG	6828	19132	1956	6658	5972	5287	8883	9923	3024	-
NDF	136	141	127	166	165	163	163	162	161	-
NDG	567	1130	383	1569	1513	1259	2047	2132	761	-
PNS	36.7	50.0	70.0	0	0	3.3	53.3	66.7	20.0	100.0
FFV	.54E-6	.51E-1	.11E-10	-	-	.16E-3	.12E-10	.42E-7	.0	.71E-7
FVC	.42E-2	.70E-3	.26E+1	-	-	.53E-2	.45E-1	.60E-5	.59E+3	.13E-1
F	0	0	0	0	0	0	0	0	0	0
PGS	57.9	20.0	33.3	6.7	0	10.3	42.9	0	0	0
GFV	-.65E+1	-.20E+1	-.71E+1	-.12E-2	-	-.17E-1	-.38E-2	0	0	0
GVC	.71E-5	.15E-6	.41E-3	.19E-3	-	.30E-6	.12E-3	-	-	-

Table 71: Test results for FMIN(1) (general problems).

Class	1B	2B	3B	4B	5B	6B	7B	8B	9B	10B
FV	.51E-3	.58E-3	.63E-3	.41E-3	.44E-3	.48E-3	.53E-3	.68E-4	.83E-4	.83E-4
VC	.14E-9	.82E-11	.82E-11	.89E-11	.11E-10	.10E-10	.61E-11	.99E-8	.32E-6	.50E-6
KT	.35E-1	.32E-1	.31E-1	.33E-1	.23E-1	.20E-1	.21E-1	.33E-2	.36E-2	.30E-2
ED	2.51	2.52	2.51	2.47	2.85	2.70	2.41	4.80	4.68	4.79
A	4.28	4.58	4.57	4.60	4.71	4.67	4.65	4.86	4.42	4.42
ET	77.9	78.4	79.4	75.9	76.6	76.0	76.3	80.3	78.0	77.5
NF	456	460	464	461	471	468	471	483	475	477
NG	4570	4607	4650	4612	4714	4685	4719	4855	4767	4786
NDF	111	111	111	111	112	111	111	111	111	111
NDG	1090	1087	1087	1086	1094	1092	1094	1073	1065	1072
ET/A	19.0	17.4	17.6	16.8	16.6	16.6	16.7	17.1	18.1	18.0
NF/A	112.6	102.4	103.4	102.6	102.5	102.5	103.5	102.7	109.8	110.3
NG/A	1127.1	1024.1	1034.9	1026.6	1025.9	1025.4	1035.7	1030.6	1102.2	1107.4
NDF/A	27.3	24.8	24.9	24.8	24.4	24.5	24.4	23.8	25.8	25.9
NDG/A	265.7	239.5	239.9	238.9	235.5	236.4	237.9	229.5	247.9	249.8
R	0	0	0	0	0	0	0	0	0	0

<u>Table 72:</u> Test results for FMIN(1) (degenerate, ill-conditioned, and indefinite problems).

Class	1A	2A	3A	4A	5A	6A	7A	8A	9A	10A
FV	.37E-4	.52E-3	.10E-4	.91E-4	.18E-3	.54E-4	.44E-4	-	.67E-4	-
VC	.47E-4	.15E-3	.12E-3	.52E-6	.31E-7	.23E-6	.25E-3	-	.20E-5	-
KT	.49E-2	.76E-4	.50E-2	.32E-2	.63E-2	.69E-2	.88E-2	-	.45E-2	-
ED	3.59	5.88	3.16	2.83	2.61	2.83	2.45	-	2.69	-
ET	71.4	235.5	222.5	181.7	151.8	59.6	428.5	-	506.1	-
NF	1525	2187	2053	2027	2205	2434	2703	-	5020	-
NG	8061	21587	6160	19339	20410	18906	34964	-	23396	-
NDF	0	0	0	0	0	0	0	-	0	-
NDG	0	0	0	0	0	0	0	-	0	-
PNS	53.3	73.3	80.0	6.7	13.3	10.0	93.3	100.0	73.3	100.0
FFV	.11E-5	.33E-1	.81E-11	.40E-7	.91E-3	.16E-2	.51E-9	.21E-3	.79E-5	.84E-8
FVC	.46E-1	.49E-4	.15E+1	.40E-6	.70E-6	.30E-3	.16E-1	.29E-2	.23E-1	.47E-1
F	0	0	0	0	0	0	0	0	0	0
PGS	50.0	25.0	0	0	0	11.1	0	0	0	0
GFV	-.73E+1	-.20E+1	-	-	-	-.17E-1	-	-	-	-
GVC	.24E-3	.16E-7	-	-	-	.23E-9	-	-	-	-

Table 73: Test results for FMIN(2) (general problems).

Class	1B	2B	3B	4B	5B	6B	7B	8B	9B	10B
FV	.22E-4	.80E-5	.32E-4	.16E-4	.41E-3	.21E-3	.12E-3	.57E-3	.61E-3	.63E-3
VC	.30E-9	.19E-9	.23E-10	.35E-10	.53E-9	.28E-10	.11E-9	.50E-9	.48E-9	.30E-4
KT	.34E-2	.37E-2	.68E-2	.45E-2	.22E-1	.12E-1	.86E-2	.21E-1	.21E-1	.22E-1
ED	3.26	3.23	3.03	2.91	2.83	2.90	2.55	4.05	4.03	3.92
A	4.97	5.12	5.08	5.13	4.29	4.77	4.62	4.57	4.56	3.33
ET	282.1	253.3	205.1	197.1	116.9	139.1	140.7	127.0	127.2	124.8
NF	3350	3094	2474	2365	1409	1663	1660	1497	1506	1480
NG	33055	30432	24269	23196	13778	16326	16419	14474	14467	14312
NDF	0	0	0	0	0	0	0	0	0	0
NDG	0	0	0	0	0	0	0	0	0	0
ET/A	58.6	50.2	41.2	39.3	27.5	28.9	30.1	29.2	29.1	37.5
NF/A	695.3	613.3	496.9	471.9	331.8	346.5	355.8	344.9	346.4	445.4
NG/A	6875.4	6023.9	4888.9	4634.5	3242.1	3391.3	3506.3	3318.0	3306.2	4302.7
NDF/A	0.0	0.0	0.0	0.0	0.0	0.0	0.0	0.0	0.0	0.0
NDG/A	0.0	0.0	0.0	0.0	0.0	0.0	0.0	0.0	0.0	0.0
R	0	0	0	0	0	0	0	0	0	0

Table 74: Test results for FMIN(2) (degenerate, ill-conditioned, and indefinite problems).

Class	1A	2A	3A	4A	5A	6A	7A	8A	9A	10A
FV	.33E-5	.34E-5	.98E-7	.18E-7	.19E-7	.91E-8	–	.28E-6	–	.15E-3
VC	.97E-6	.52E-5	.22E-6	.10E-7	.29E-7	.20E-8	–	.52E-7	–	.29E-4
KT	.75E-3	.18E-5	.16E-5	.65E-4	.11E-3	.21E-5	–	.30E-3	–	.17E-1
ED	4.16	8.05	6.84	5.32	5.01	6.31	–	4.72	–	3.02
ET	54.1	145.8	206.8	178.9	217.2	44.2	–	530.2	–	260.9
NF	479	627	235	275	270	284	–	448	–	239
NG	3033	7715	705	2771	2730	2309	–	6421	–	4501
NDF	426	360	336	533	598	680	–	1169	–	1011
NDG	1723	3087	1159	5080	5559	5275	–	15550	–	17315
PNS	33.3	63.3	27.8	0	0	0	100.0	44.4	–	13.3
FFV	.23E-6	.58E-3	.0	–	–	–	.0	.42E-4	–	.29E-2
FVC	.25E+2	.62	.23E+1	–	–	–	.87	.85E-2	–	.99E-3
F	0	0	4	0	0	0	4	2	5	0
PGS	40.0	18.2	53.8	0	0	10.0	0	0	–	0
GFV	-.53E+1	-.20E+1	-.13E+2	–	–	-.17E-1	–	–	–	–
GVC	.11E-4	.38E-4	.88E-5	–	–	.18E-6	–	–	–	–

Table 75: Test results for FMIN(3) (genral problems).

Class	1B	2B	3B	4B	5B	6B	7B	8B	9B	10B
FV	.67E-7	.80E-7	.53E-7	.79E-7	.11E-6	.26E-6	.96E-8	.52E-8	.52E-8	.59E-8
VC	.38E-8	.13E-9	.78E-10	.0	.15E-7	.72E-8	.16E-8	.57E-9	.22E-9	.36E-9
KT	.14E-3	.29E-3	.26E-3	.30E-3	.48E-3	.13E-3	.24E-5	.83E-6	.56E-6	.76E-6
ED	4.82	4.22	4.36	3.96	4.51	4.56	3.89	7.78	7.92	7.73
A	6.06	6.19	6.33	6.65	5.64	5.79	6.58	7.85	8.03	7.88
ET	232.4	231.4	234.1	233.7	192.1	205.9	247.1	197.0	197.9	195.9
NF	228	234	238	228	198	219	252	203	202	205
NG	2309	2356	2407	2301	2005	2214	2536	2051	2037	2067
NDF	538	552	563	534	461	491	581	449	450	506
NDG	5305	5423	5524	5242	4516	4838	5769	4381	4377	4370
ET/A	38.3	37.4	36.9	35.2	33.2	35.1	37.8	25.6	25.0	25.3
NF/A	37.7	37.9	37.8	34.4	34.3	37.1	38.4	26.4	25.4	26.3
NG/A	381.0	381.9	381.3	346.5	346.1	374.8	386.4	265.9	256.2	265.4
NDF/A	88.9	89.6	89.1	80.4	79.5	83.5	88.9	58.0	56.7	57.4
NDG/A	875.0	876.8	872.6	789.7	779.8	822.1	882.9	565.1	548.4	559.9
R	0	0	0	0	0	0	0	0	0	0

Table 76: Test results for FMIN(3) (degenerate, ill-conditioned, and indefinite problems).

Class	1A	2A	3A	4A	5A	6A	7A	8A	9A	10A
FV	.22E-5	–	.18E-5	.74E-6	.11E-5	.19E-5	–	–	.38E-6	–
VC	.18E-5	–	.24E-5	.16E-5	.22E-5	.23E-5	–	–	.85E-6	–
KT	.26E-3	–	.22E-5	.13E-3	.17E-3	.15E-3	–	–	.25E-4	–
ED	5.03	–	6.47	4.43	4.49	4.46	–	–	3.43	–
ET	134.2	–	134.7	88.5	92.1	44.4	–	–	183.8	–
NF	2735	–	1844	694	784	1500	–	–	868	–
NG	16414	–	5532	6942	7846	12005	–	–	4343	–
NDF	212	–	106	82	94	163	–	–	76	–
NDG	1278	–	319	825	945	1306	–	–	381	–
PNS	46.7	–	44.4	3.3	0	0	–	–	0	–
FFV	.44E-5	–	.0	.17E-2	–	–	–	–	–	–
FVC	.37	–	.67E+3	.14E-5	–	–	–	–	–	–
F	0	10	4	0	0	0	5	5	2	5
PGS	81.3	–	80.0	0	0	6.7	–	–	0	–
GFV	-.85	–	-.13E+2	–	–	-.17E-1	–	–	–	–
GVC	.13E-5	–	.32E-5	–	–	.21E-5	–	–	.	–

Table 77: Test results for NLP (general problems).

Class	1B	2B	3B	4B	5B	6B	7B	8B	9B	10B
FV	.22E-5	.60E-6	.10E-5	.24E-9	.95E-6	.12E-5	.14E-5	.27E-5	.16E-5	.23E-5
VC	.23E-5	.14E-5	.17E-5	.20E-6	.21E-5	.24E-5	.28E-5	.36E-5	.35E-5	.35E-5
KT	.42E-3	.77E-4	.16E-3	.99E-5	.33E-3	.14E-3	.39E-4	.16E-4	.54E-5	.13E-4
ED	4.29	4.34	4.24	4.95	4.57	4.34	2.73	5.24	5.69	5.44
A	4.74	5.14	4.95	6.57	4.94	4.94	4.64	5.27	5.56	5.36
ET	112.3	81.9	86.7	37.3	92.6	111.2	143.9	180.6	177.4	170.3
NF	1025	732	770	286	827	1030	1392	1758	1720	1588
NG	10252	7321	7707	2860	8276	10305	13920	17585	17208	15881
NDF	100	76	83	43	86	99	112	129	133	126
NDG	1005	760	831	432	866	990	1121	1292	1336	1266
ET/A	23.6	16.1	17.6	6.1	18.7	22.6	31.0	34.9	32.4	32.6
NF/A	215.4	143.8	156.3	46.9	167.1	209.4	299.8	341.9	314.8	306.3
NG/A	2154.4	1437.9	1562.6	469.2	1670.5	2093.5	2998.2	3419.2	3147.6	3062.6
NDF/A	21.2	14.9	16.8	7.0	17.5	20.2	24.2	25.0	24.4	24.1
NDG/A	212.0	148.5	168.3	69.9	175.3	201.7	241.8	249.8	243.5	241.4
R	0	0	0	0	0	0	0	0	0	0

Table 78: Test results for NLP (degenerate, ill-conditioned, and indefinite problems).

Class	1A	2A	3A	4A	5A	6A	7A	8A	9A	10A
FV	.47E-4	–	.29E-3	.27E-5	.27E-6	.56E-3	.40E-4	.11E-5	.34E-5	.26E-5
VC	.13E-5	–	.42E-3	.0	.0	.0	.31E-5	.0	.28E-6	.0
KT	.25E-2	–	.53E-2	.11E-2	.33E-3	.31E-1	.18E-2	.49E-3	.10E-2	.13E-2
ED	3.72	–	3.37	3.93	4.52	2.28	3.41	4.62	3.17	4.33
ET	107.3	–	183.2	211.5	232.4	65.7	302.3	361.8	453.2	602.7
NF	1740	–	1474	1915	2513	2125	1391	2535	2909	3963
NG	11750	–	4419	20540	26345	18983	18555	37978	15249	77788
NDF	75	–	56	79	102	95	70	130	117	208
NDG	498	–	168	859	1063	840	921	1918	603	4014
PNS	79.2	100.0	62.5	63.3	50.0	66.7	93.3	66.7	60.0	66.7
FFV	.16E-5	.14E+3	.12E-9	.11	.42E-1	.18	.47E-6	.24	.34E-3	.18E+1
FVC	.20E+1	.0	.84E-2	.33E-9	.76E-10	.34E-6	.55E-1	.0	.31E-2	.0
F	2	3	2	0	0	8	0	1	0	3
PNG	40.0	0	66.7	0	0	50.0	0	0	0	0
GFV	-.12E+1	–	-.16	–	–	-.17E-1	–	–	–	–
GVC	.10E-3	–	.22E-3	–	–	.0	–	–	–	–

Table 79: Test results for SUMT (general problems).

Class	1B	2B	3B	4B	5B	6B	7B	8B	9B	10B
FV	.32E-4	.31E-4	.31E-4	.16E-4	.37E-5	.26E-4	.41E-4	.22E-4	.25E-4	.21E-4
VC	.0	.0	.0	.0	.0	.0	.0	.0	.0	.0
KT	.34E-2	.50E-2	.37E-2	.48E-2	.11E-2	.32E-2	.11E-2	.75E-4	.89E-4	.25E-4
ED	3.30	3.05	3.08	2.95	4.05	3.26	2.69	5.06	5.09	5.13
A	5.57	5.47	5.51	5.51	6.11	5.58	5.51	6.46	6.43	6.60
ET	101.5	101.0	101.0	101.4	100.6	101.8	101.4	96.5	97.9	99.4
NF	1019	1042	1030	1036	1042	1052	1029	918	938	950
NG	10398	10608	10495	10559	10659	10725	10490	9965	10188	10314
NDF	53	54	54	54	51	54	54	48	48	49
NDG	536	550	552	550	520	546	548	516	508	522
ET/A	18.3	18.5	18.3	18.4	16.5	18.2	18.4	15.1	15.3	15.3
NF/A	183.3	190.8	187.2	188.4	170.6	188.5	187.0	143.6	147.2	146.1
NG/A	1869.5	1941.8	1906.3	1918.3	1744.7	1921.2	1905.1	1556.7	1595.5	1583.3
NDF/A	9.6	10.0	9.9	9.9	8.4	9.7	9.9	7.7	7.5	7.6
NDG/A	96.5	100.6	100.4	99.8	85.1	97.8	99.6	80.9	79.7	80.4
R	0	0	0	0	0	0	0	0	0	0

Table 80: Test results for SUMT (degenerate, ill-conditioned, and indefinite problems).

Class	1A	2A	3A	4A	5A	6A	7A	8A	9A	10A
FV	.36E-6	.25E-3	.97E-8	.28E-6	.20E-8	-	-	.57E-7	.37E-4	-
VC	.15E-6	.50E-3	.21E-7	.37E-10	.37E-9	-	-	.35E-10	.23E-3	-
KT	.72E-4	.88E-5	.52E-5	.44E-3	.57E-4	-	-	.17E-3	.20E-2	-
ED	5.37	6.78	6.56	4.66	5.67	-	-	5.08	2.76	-
ET	83.1	225.1	84.0	50.1	54.6	-	-	242.7	184.0	-
NF	1172	1355	422	334	382	-	-	1258	582	-
NG	7034	16265	1266	3344	3822	-	-	17613	2913	-
NDF	170	160	72	50	57	-	-	173	86	-
NDG	1025	1921	217	507	573	-	-	2429	431	-
PNS	14.3	42.9	4.8	29.2	56.7	-	-	26.7	50.0	-
FFV	.28E+1	.60E-5	.65E-2	.22E+33	.19E-1	-	-	.23E+10	.28E+18	-
FVC	.17E+1	.53E+1	.43E-6	.59E+14	.41E-6	-	-	.11E+6	.98E+25	-
F	3	3	3	2	0	10	5	0	3	5
PGS	55.6	0	75.0	0	0	-	-	0	0	-
GFV	-.99	-	-.31E+1	-	-	-	-	-	-	-
GVC	.30E-6	-	.18E-4	-	-	-	-	-	-	-

Table 81: Test results for DFP (general problems).

Class	1B	2B	3B	4B	5B	6B	7B	8B	9B	10B
FV	.90E-4	.14E-4	.16E-3	.17E-9	.10E-3	.23E-2	.57E-3	.94E-8	.15E-8	.30E-8
VC	.10E-7	.11E-3	.13E-8	.34E-6	.0	.34E-12	.68E-3	.68E-11	.15E-10	.13E-10
KT	.14E-1	36E-2	.13E-1	.84E-6	.12E-1	.59E-1	.37E-2	.37E-4	.24E-4	.14E-4
ED	2.98	3.11	2.64	4.84	3.38	2.34	1.97	7.18	7.79	7.75
A	4.21	3.58	4.30	6.79	5.08	4.67	2.70	7.70	8.02	8.01
ET	68.3	51.4	118.6	38.2	36.8	112.9	199.4	42.9	47.7	43.2
NF	510	362	899	259	281	954	1755	297	329	297
NG	5100	3624	8994	2592	2810	9540	17550	2971	3293	2977
NDF	66	52	117	44	40	109	184	45	49	43
NDG	664	528	1172	440	400	1090	1840	453	491	435
ET/A	17.3	15.4	25.3	6.0	7.2	24.2	73.9	5.6	6.0	5.4
NF/A	129.8	108.6	190.7	40.5	55.3	204.3	650.0	38.7	41.4	37.5
NG/A	1297.5	1086.0	1907.1	404.9	553.1	2042.8	6500.0	387.0	414.3	374.9
NDF/A	16.9	15.9	25.0	6.8	7.9	23.3	68.1	5.9	6.2	5.5
NDG/A	169.4	159.0	249.9	68.4	78.7	233.4	681.4	59.2	61.8	54.6
R	1	2	2	0	1	2	7	0	0	0

Table 82: Test results for DFP (degenerate, ill-conditioned, and indefinite problems)

Class	1A	2A	3A	4A	5A	6A	7A	8A	9A	10A
FV	.65E-4	.23E-5	.23E-3	.11E-5	.34E-5	.72E-7	.96E-4	.37E-4	.18E-4	.59E-5
VC	.95E-4	.59E-8	.28E-3	.10E-11	.0	.48E-9	.20E-3	.35E-10	.64E-4	.74E-5
KT	.38E-4	.15E-6	.44E-4	.33E-3	.64E-3	.48E-4	.11E-2	.27E-2	.15E-3	.89E-3
ED	5.30	9.31	5.35	4.38	4.05	5.59	3.36	3.87	3.34	4.90
ET	16.2	11.0	31.1	9.3	14.5	13.3	100.9	18.3	68.5	45.6
NF	208	57	76	78	107	208	199	61	159	160
NG	429	353	204	195	306	513	1234	438	263	1657
NDF	94	28	43	40	53	107	105	30	81	85
NDG	224	183	117	97	147	276	669	217	139	905
PNS	56.7	66.7	20.0	20.0	6.7	16.7	13.3	0	0	0
FPV	.33E-5	.21E-7	.30	.32E+1	.25E+1	.84	.81E-6	-	-	-
FVC	.16E+2	.30E+3	.27E-1	.15E-1	.10E-1	.25	.77E-1	-	-	-
F	0	1	0	0	0	0	0	0	1	0
PGS	69.2	22.2	95.8	0	0	12.0	84.6	0	0	20.0
GFV	-.22E+2	-.20E+1	-.65E+3	-	-	-.17E-1	-.51E-1	-	-	-.17E-2
GVC	.12E-3	.46E-9	.30E-3	-	-	.23E-9	.21E-3	-	-	.63E-4

Table 83: Test results for FCDPAK (general problems).

Class	1B	2B	3B	4B	5B	6B	7B	8B	9B	10B
FV	.52E-5	.20E-5	.27E-4	.84E-6	.30E-5	.20E-4	.27E-4	.15E-4	.15E-4	.15E-4
VC	.0	.0	.0	.0	.0	.0	.0	.0	.0	.0
KT	.14E-2	.64E-3	.29E-2	.83E-3	.13E-2	.20E-2	.20E-3	.17E-3	.18E-3	.13E-3
ED	3.71	3.88	3.18	3.48	4.03	3.35	2.65	4.85	4.90	4.88
A	5.96	6.19	5.57	6.16	6.11	5.69	5.73	6.36	6.36	6.40
ET	6.3	5.3	5.0	4.2	5.4	4.4	4.3	3.1	3.0	3.1
NF	44	36	39	26	36	32	23	13	12	12
NG	116	88	75	66	94	64	70	49	50	50
NDF	22	19	19	13	18	16	11	8	7	7
NDG	54	42	34	30	44	29	33	24	25	26
ET/A	1.1	.9	.9	.7	.9	.8	.8	.5	.5	.5
NF/A	7.5	6.0	7.2	4.2	6.0	5.7	4.1	2.2	2.0	2.0
NG/A	19.4	14.1	13.8	10.5	15.3	11.4	12.3	7.8	7.9	7.9
NDF/A	3.8	2.9	3.5	2.1	3.0	2.9	2.1	1.3	1.2	1.2
NDG/A	9.1	6.7	6.3	4.9	7.1	5.3	5.8	3.9	4.0	4.1
R	0	0	0	0	0	0	0	0	0	0

Table 84: Test results for FCDPAK (degenerate, ill-conditioned, and indefinite problems).